Further praise for *Symphony*

"Covering topics from carbon's ancient origins to the threats that carbon compounds pose to our future climate, [Robert M.] Hazen's book is a fascinating read. *Symphony in C* chronicles cutting-edge science that's helping researchers make better sense of the carbon-rich world around us."   —Sid Perkins, *Science News*

"Hazen's enthusiasm, the string of shareable facts presented, and the introduction of so many interesting scientists . . . make this book such a fascinating read. . . . Hazen brings a distinct and intentionally personal perspective to this topic. . . . Throughout *Symphony in C*, science is presented as a living and very human endeavor."
                                                           —Nicola Pohl, *Science*

"Hazen sets the record straight in this thoughtful love letter to [carbon]."   —Gemma Tarlach, *Discover*

"[A] lively, expert overview. . . . Hazen [is] a smooth stylist. . . . A skillful account of the central element in our lives."
                                                           —*Kirkus Reviews*

"Hazen brings the process of scientific investigation to life. . . . [He] conveys the delight he finds in the process of understanding the world around him. . . . [This] enthusiastic survey also shows the limits of existing knowledge and the potential for future discoveries in an exciting field."   —*Publishers Weekly*

"Probing. . . . Science that burrows into issues of profound interest."
                                                           —*Booklist*

"From the Big Bang to coal, carbohydrates, and ultra-strong high-tech nanofibers, Robert M. Hazen provides an illuminating and enjoyable guide to the remarkable odyssey of carbon, the element of life. Enjoy the trip!"   —Andrew Knoll, Fisher Professor of Natural History, Harvard University

"Robert M. Hazen's virtuoso performance captures the wonder of the sixth element—from volcanic gases to al dente pasta to life's very beginnings—while telling the wonderful stories about the people behind the discoveries."

—Terry Plank, Arthur D. Storke Memorial Professor of Geochemistry, Columbia University

"This book is an incredibly rich story of carbon and its role in life. Robert M. Hazen has outdone himself in delivering an engaging, edifying, great read. If you don't know why carbon is important in your life, or even if you think you do, you should put down whatever you're reading and get this book."

—Paul G. Falkowski, author of *Life's Engines*

Symphony in C

Also by Robert M. Hazen

The Story of Earth:
The First 4.5 Billion Years, from Stardust to Living Planet

Genesis: The Scientific Quest for Life's Origin

The Diamond Makers

The New Alchemists:
Breaking through the Barriers of High-Pressure Research

The Sciences: An Integrated Approach (with James Trefil)

Why Aren't Black Holes Black: The Unanswered Questions
at the Frontiers of Science (with Maxine Singer)

The Breakthrough: The Race for the Superconductor

Science Matters: Achieving Scientific Literacy (with James Trefil)

Keepers of the Flame:
The Role of Fire in American Culture, 1775–1925 (with Margaret Hazen)

Wealth Inexhaustible: An Introduction to the History of
American Mineral Industries to 1850 (with Margaret Hazen)

The Music Men: An Illustrated History of
American Brass Bands, 1880–1920 (with Margaret Hazen)

Comparative Crystal Chemistry (with Larry Finger)

North American Geology

The Poetry of Geology

American Geological Literature (with Margaret Hazen)

Symphony in C

Carbon and the Evolution of (Almost) Everything

Robert M. Hazen

W. W. NORTON & COMPANY
Independent Publishers Since 1923

For information about permission to reproduce selections from
this book, write to Permissions, W. W. Norton & Company, Inc.,
500 Fifth Avenue, New York, NY 10110

For information about special discounts for bulk purchases, please contact
W. W. Norton Special Sales at specialsales@wwnorton.com or 800-233-4830

Manufacturing by LSC Harrisonburg
Book design by Patrice Sheridan
Production manager: Julia Druskin

Library of Congress Cataloging-in-Publication Data

Names: Hazen, Robert M., 1948– author.
Title: Symphony in C: carbon and the evolution of (almost) everything /
Robert M. Hazen.
Other titles: Carbon and the evolution of (almost) everything
Description: First edition. | New York: W.W. Norton & Company, [2019] | Includes
bibliographical references and index.
Identifiers: LCCN 2019013772 | ISBN 9780393609431 (hardcover)
Subjects: LCSH: Carbon. | Carbon cycle (Biogeochemistry) | Chemical bonds. |
Chemistry, Organic.
Classification: LCC QD181.C1 H39 2019 | DDC 577/.144—dc23
LC record available at https://lccn.loc.gov/2019013772

ISBN 978-0-393-35862-9 pbk.

W. W. Norton & Company, Inc., 500 Fifth Avenue, New York, N.Y. 10110
www.wwnorton.com

W. W. Norton & Company Ltd., 15 Carlisle Street, London W1D 3BS

1 2 3 4 5 6 7 8 9 0

For my friends and colleagues
of the Deep Carbon Observatory

The adventure has only just begun.

Contents

Prologue

LOOK AROUND YOU. Carbon is everywhere: in the paper of this book, the ink on its pages, and the glue that binds it; in the soles and leather of your shoes, the synthetic fibers and colorful dyes of your clothes, and the Teflon zippers and Velcro strips that fasten them; in every bite of food you eat, in beer and booze, in fizzy water and sparkling wine; in the carpets on your floors, the paint on your walls, and the tiles on your ceilings; in fuels from natural gas to gasoline to candle wax; in sturdy wood and polished marble; in every adhesive and every lubricant; in the lead of pencils and the diamond of rings; in aspirin and nicotine, codeine and caffeine, and every other drug you've ever taken; in every plastic, from grocery bags to bicycle helmets, cheap furniture to designer sunglasses. From your first baby clothes to your silk-lined coffin, carbon atoms surround you.

Carbon is the giver of life: Your skin and hair, blood and bone, muscle and sinews all depend on carbon. Every cell in your body—indeed, every part of every cell—relies on a sturdy backbone of carbon. The carbon of a mother's milk becomes the carbon of her child's beating heart. Carbon is the chemical essence of your lover's eyes, hands, lips, and brain. When you breathe, you exhale carbon; when you kiss, carbon atoms embrace.

It would be easier for you to list everything you touch that lacks carbon—aluminum cans in your fridge, silicon microchips in your iPhone, gold fillings in your teeth, other oddities—than to enumerate even 10 percent of the carbon-bearing objects in your life. We live on a carbon planet and we are carbon life.

Every chemical element is special, but some elements are more special than others. Of all the periodic table's richly varied denizens, the sixth element is unique in its impact on our lives. Carbon is not simply the static element of "stuff." Carbon provides the most critical chemical link across the vastness of space and time—the key to understanding cosmic evolution. Over the course of almost 14 billion years, the Universe has evolved and become ever more richly patterned, with seemingly endless fascinating and quirky behaviors. Carbon lies at the heart of this evolution—choreographing the emergence of planets, life, and us. And, more than any other ingredient, carbon has facilitated the rapid emergence of new technologies, from steam engines of the Industrial Revolution to our modern "Plastic Age," even as it accelerates unprecedented changes in environment and climate on a planetary scale.

Why focus on carbon? Hydrogen is a far more abundant chemical element, helium more stable, and oxygen more reactive. Iron, sulfur, phosphorus, sodium, calcium, nitrogen—all have fascinating stories to tell. All played critical roles in Earth's complex evolution. But if you wish to find meaning and purpose in the vast cold and dark of the Universe, look to carbon; carbon, by itself and in chemical combinations with other atoms, provides unmatched cosmic novelty and unparalleled potential for cosmic evolution.

Of more than 100 chemical elements, carbon stands out as an element of our aspirations and fears. Novel carbon-based materials, invented by the thousands every year—Kleenex, spandex, Freon, nylon, polyethylene, Vaseline, Listerine, Bactine, Scotch tape, Silly Putty—enhance our lives in countless ways, both seen and unseen. But the proliferation of these synthetic chemicals has led to unintended consequences: troubling holes in the protective ozone layer, deadly allergic reactions, and carcinogens by the score. As the basis of all biomolecules, no other element contributes so centrally to the well-being and sustainability of life on Earth, including our human species. But carbon atoms, when missing or misaligned, can lead to disease and death.

The near-surface carbon cycle stabilizes Earth's climate, ensures the health of ecosystems, and provides us with our most abundant supplies of inexpensive energy. Yet, if the distribution of carbon atoms becomes skewed by natural or human activities—erupting volcanoes, burning coal, an errant asteroid, vanishing forests—climates can change and ecosystems can collapse. And carbon's influence is not confined to the near-surface realm of the living; carbon's behavior in Earth's hidden, deep interior epitomizes the dynamic processes that set apart our planet from all other known worlds.

The story of carbon is, in a sense, the story of everything. Yet mysteries about this ubiquitous, indispensable element abound. We don't know how much carbon Earth holds, nor do we fully comprehend its varied forms hidden deep within our planet. We don't understand the movements of carbon atoms as they cycle between Earth's surface and its deep interior, nor can we say whether those movements have changed significantly through billions of years of Earth history—through "deep time." Despite the existence of millions of known carbon compounds, scientists have only just begun to explore the richness of carbon chemistry. And the greatest mystery of all—the origin of life—is inextricably linked to the behavior of carbon in complex chemical combinations with other elements.

From quantities and forms, to movements and origins, what we know about carbon is dwarfed by our ignorance. We must find answers, but how can we hope to bridge such yawning chasms in our understanding? The very structure of the scientific enterprise would seem to conspire against sustained progress. Universities lack departments of carbon science, and large-scale, cross-disciplinary research ventures are rare. Scientific discovery rests on asking questions about the natural world, but it also depends on finding resources in a climate of limited time and money, at a time when disciplinary specialization often trumps integration.

Who will champion a different kind of research support?

—

The scene is the venerable Century Association club in New York City, early 2007, where the fund-raisers with the Carnegie Institution for Science have invited several dozen potential donors to an elegant dinner. The economy is booming, and Barack Obama is still a senator from Illinois. Paintings and sculptures by some of the greatest artists in American history line the spacious, wood-paneled rooms of the club. The major artworks, by such luminaries as John Frederick Kensett, Winslow Homer, and Paul Manship, were proffered in exchange for coveted and pricey club memberships. It was a great deal all around: the Century Association built a superb collection of masterpieces, while the artists gained access to wealthy patrons who could afford the club's steep initiation fees.

I was the after-dinner speaker, and my theme was origins-of-life research, an intrinsically entertaining topic that was enhanced by simple props: a glass of carbonated soda water, a rock picked up in a nearby park, a teaspoon, and a straw. Presto!—a user-friendly demonstration of the chemical steps by which life might have emerged from a deep, hot, carbon-rich volcanic environment on the ocean floor. That my ideas were a bit controversial—a source of a lively, sometimes acrimonious debate with skeptical peers—added a bit of spice to my remarks. As a bonus, everyone was given a copy of *Genesis*, my recent book on the subject. I remember feeling a kinship with the artists whose work hung about me. Like them, I was singing for my supper, trying to catch the eye of some prospective patron, hoping for that next commission that would allow my colleagues and me to create a new scientific canvas.

Science isn't cheap. It can cost $100,000 per year to support each graduate student or postdoc. New analytical machines can run upwards of a million dollars, with service contracts and replacement parts adding 10 percent or more per year to the price tag. Travel to conferences, page charges for publications, and basic lab supplies like test tubes, reagents, and Kimwipes are essential. And don't get me started on "overhead." Without support from industry, government agencies, and private foundations, scientific research would quickly wither and die. But it's a tough road writing grant proposals to agencies and foundations, requesting $100,000 per year with less than a 10 percent chance of winning funding.

So there I was in the Big Apple, hat in hand, promoting science to a room full of nonscientists. One might do a score of such events without a nibble, but you have to try. The evening was fun, but soon forgotten in the crush of research projects and grant deadlines. And then came the phone call that changed everything.

—

It was three months later, early spring 2007, as Washington was coming into bloom.

"Hi Bob. Jesse Ausubel here, from the Sloan Foundation in New York." Apparently I'd met Jesse at the Century Association talk, but I didn't remember him. He seemed cordial but businesslike, his voice a pleasant baritone.

"The Sloan Foundation is considering new programs." My ears perked. Sloan supports major science research and education efforts: an ambitious Census of Marine Life, the digital sky survey that discovered dark energy, NPR, and PBS.

"We're wondering whether you'd be interested in discussing a program on the deep origins of life?" The subject of my New York talk, the very speculative hypothesis that life emerged from a deep volcanic zone on the ocean floor, had evidently hit the mark.

Ausubel told me that Sloan's programs typically run for ten years at $7 to $10 million per year, paused, and waited for some kind of reaction. From me, silence. A one with eight zeros paralyzed my brain.

Eventually, I recovered and we began to discuss details. I suggested that focusing exclusively on deep origins of life was too narrow a view for a big ten-year effort. A host of fundamental mysteries relate to carbon at the planetary scale—not just in biology, but also in physics, chemistry, and geology. I explained that we can't really understand life's ancient, mysterious origins until we understand the broader story of carbon in Earth.

Jesse Ausubel liked the idea of a comprehensive approach: physics, chemistry, geology, and biology; 4.5 billion years of Earth history; crust to core, at scales from nano to global. He offered a one-year, $400,000 exploratory grant—"preapproved," he said—to gather experts from around the world, hold workshops, define what we know and what we don't know, and consider a global strategy to transform our understanding of Earth's carbon.

This was no mere canvas. It was an insanely ambitious Beethoven symphony with unprecedented forces—a great bellowing chorus, multiple operatic soloists, and an oversized orchestra with myriad voices from tuba to piccolo. Nothing like it had ever been attempted before.

—

Fast-forward a year, to May 15, 2008. More than 100 experts gathered from around the world.[1] Distinguished senior professors joined early career scientists from a dozen countries and as many scientific disciplines. We were tasked with discovering whether the rationale and will existed to tackle carbon science in a new, integrated approach.

Day 1 wasn't all that encouraging, as scientists seldom stray far from their comfort zones. In spite of lofty rhetoric about "abandoning silos" and "crossing boundaries," the biologists pretty much talked to biologists, while geophysicists and organic chemists also huddled within their specialized subgroups.

Day 2 was better. Gradually, as a succession of vivid talks provided glimpses of unexplored vistas—the puzzle of carbon in Earth's core, the enigmatic ancient origins of life, the stately cycling of plate tectonics, hints of a vast subsurface microbial biosphere—we saw our narrow specialties in new, broader contexts. For the first time, we learned

about paradoxical, unexplored connections between exploding volcanoes and diamond deposits, plate tectonics and climate change, and
chemically reactive minerals and hidden deep life. The fascination of
carbon science as a universal integrating theme seduced us.

By the end of Day 3, the structure for a new global endeavor had
been framed. Leaders emerged and enthusiasm was high. Observers
from the Sloan Foundation felt the energy in the room and saw the
commitment in our eyes; they quickly gave a green light for the Deep
Carbon Observatory.[2] Ours would be a global endeavor of unusual
scientific ambition and scope. The prospect was thrilling, but I suspect
that every participant also worried about being part of a spectacular,
embarrassing, expensive failure.

—

A decade later, the adventure has exceeded our most ambitious vision.
An international army of carbon researchers—more than a thousand scientists from fifty countries—tackles the mysteries of carbon
in Earth. With total international funding approaching a half billion
dollars from dozens of agencies and foundations worldwide, the Deep
Carbon Observatory represents one of the most comprehensive and
broadly interdisciplinary scientific endeavors in history.

As with any successful science program, we have learned a lot, but
we've also become more keenly aware of how much we don't know.
Nagging, unanswered questions have become more deeply etched,
more insistent drivers of future research. The paradox of science is
that the more we know, the more we realize is unknown, perhaps
even unknowable. Each discovery opens a door to a vaster unexplored
landscape.

I'm driven to share some of the emerging, breathtaking vistas of
carbon science—to chronicle the discoveries made, as well as the great
unknown that remains to be explored. But how? If I were John Frederick Kensett or Winslow Homer, perhaps I could paint a picture. Words
are harder. A multivolume encyclopedia of carbon could scarcely do

justice to the many nuances of the subject. How, then, can carbon's story be captured within the pages of a single book? The opportunity beckoned, but I was stymied. The blank page mocked me until Jesse Ausubel suggested a path forward.

"You must write a symphony!" he commanded.[3]

Jesse knew I had spent forty years as a symphony musician, juggling long days of lab work and evening gigs as a trumpeter with many groups—a regular with the Washington Chamber Symphony and National Gallery Orchestra, an extra with the National Symphony Orchestra and Washington National Opera. I had played every symphony of Beethoven, Brahms, Schuman, and Mendelssohn many times over. Still, at first his remark was puzzling. A symphony in words, not music? Four movements of . . . what?

I was uncertain and confused, but the metaphor also made sense on several levels. Like the varied physicists, chemists, biologists, and geologists of the Deep Carbon Observatory, a symphony orchestra features diverse specialists, each with years of training and dedication. Each orchestra musician has a distinctive instrument; violin and tuba, flute and snare drum, trumpet and viola—every timbre and range is essential, but none alone can unleash the swelling grandeur of the whole. So it is with the symphony of carbon science. Without the many voices of the Deep Carbon Observatory, the *Symphony in C* could never be heard.

The metaphor also recognizes that beautiful solos periodically emerge from the orchestra's fabric. Our carbon symphony thus features the exceptional contributions of individual women and men of science, even as it integrates their focused research into a larger work with grander themes.

Like every symphony, this volume is a personal journey—idiosyncratic in content, limited in scope, composed from my own biased perspective, and playing out in many moods. I have benefited from the work of hundreds of colleagues, but this telling of the carbon story is inherently personal. Many other symphonies in C are waiting to be written.

—

As the parallels between the scientific endeavor and great orchestral compositions came into focus, I warmed to the idea of *Symphony in C,* though I struggled to envision a coherent framework. Then a thought: Ancient scholars postulated the existence of four elements—Earth, Air, Fire, and Water—each "essence" with a distinctive set of characteristics, each an irreducible component of the Universe, but collectively the source of all material creation. Carbon, alone among the atoms of the periodic table, displays the varied characteristics of all four classical elements, which suggest a four-movement framework for our story.

As in a symphony, the book's four movements differ in their broad themes, their moods, and their tempi. "Movement I—Earth" examines minerals and rocks, the solid crystalline foundation of our planet. The movement begins with the dawn of creation, long before the formation of planet Earth, when atoms of carbon were forged from smaller atomic bits. It shifts to the emergence and evolution of Earth's mineral wealth—a celebration of the growing diversity and exuberant beauty of crystalline carbon compounds.

The focus of "Movement II—Air" is Earth's stately carbon cycle. Carbon atoms constantly shift among reservoirs, trading places between the oceans and atmosphere, plunging into the deep interior by way of plate tectonics and venting back to the surface in the hot gases released from hundreds of active volcanoes. For millions of years, this deep carbon cycle has enjoyed a reliable balance—an equilibrium that human actions may now be altering in ways that will lead to unintended consequences. Like the slow movement of a symphony, this topic calls for a softer, gentler treatment.

Carbon's dynamic roles in energy, industry, and emerging high-tech applications demand the punchy, fast-paced Scherzo of "Movement III—Fire." Carbon is the element of "stuff"—essential materials with myriad properties that benefit every facet of our lives. Stories of scientists and musicians punctuate the Scherzo, as carbon permeates every aspect of our lives.

And finally, "Movement IV—Water" explores the origins and evolution of life. The movement opens peacefully as life emerges from Earth's primitive ocean, but it relentlessly accelerates with life's astonishing evolutionary diversifications and innovations. *Symphony in C* rushes to a unifying Finale in which the many themes of carbon science come together.

—

Settle into your seats. The lights are dimming. Our story begins at the beginning, before carbon, even before time, as the Universe is about to emerge from absolute nothingness.

Symphony in C

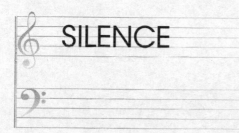

SILENCE

Before the Cosmos was the void.
Nothing—no hint of matter, or light, or even empty space.
No thought or discovery, no art or music, no hope or dream.
Darkness. Silence.

We cannot fathom such an absence of everything,
such absolute nothingness.
The time before time remains unknowable,
mysterious, beyond our catalog of physical law.
It was a time before carbon and the emergence of everything.

MOVEMENT I—EARTH

Carbon, the Element of Crystals

The moment of creation!
Time and space emerge from nothing.
The cosmic essence of everything appears in an instant
in a maelstrom of pure energy, arising from the void.

Our Universe is birthed in an outrageously concentrated form—
dense and hot and tiny beyond imagining, but it expands
faster than light, creating volume, quickly cooling as it grows.
As it cools, the Cosmos becomes more structured,
more familiar, more like home.

PRELUDE—Before Earth

CARBON'S GRAND SYMPHONY began with a brief, frenzied prelude shortly after the dawn of creation, 13.8 billion years ago. For a short interval following the Big Bang, no atoms of any kind graced the Cosmos. The Universe was too hot, too violent. The dense, superheated frenzy of matter and energy had first to expand and cool, as fundamental particles sorted themselves into the stuff of stars, planets, and life. Hydrogen and helium formed first, in giddy profusion, providing the starting point for almost every material object we know. But in a discovery only just now coming to light, prodigious numbers of heavier atoms—life-giving carbon, nitrogen, and oxygen among them—were also created.

The Invention of Atoms: Big Bang Carbon

Scientists have long taught that the story of carbon began in the stars, most likely millions of years after the Big Bang—a claim repeated in dozens of textbooks and peer-reviewed publications. That we were misled underscores key themes about the vibrant, mutable, maddening, and utterly engrossing world of carbon research. How can one avoid such pitfalls? Question every assumption, check your results and check them again, and prepare to be astonished.

Long before the first generation of stars, the only atom-making process in the history of the Universe was a unique, fleeting event—a

17-minute burst of nuclear creativity called "Big Bang nucleosynthesis," or BBN.[1] The Big Bang—that singular, enigmatic instant 13.8 billion years ago when all matter and energy and space itself suddenly came into existence at a point—began the expansion that to this day characterizes our Universe. Expansion means cooling, and with cooling came a sequence of condensations—what physicists call "freezings"—each a cascade of transformations, each of which made the Cosmos more patterned and interesting.

The first to condense out of the incomprehensibly hot and dense maelstrom were the most elementary of particles, called "quarks" (building blocks of atomic nuclei) and "leptons" (think electrons). Within the first second of time, when temperatures dropped to an unfathomable 100 trillion degrees, triplets of quarks combined into astronomical numbers of individual protons and neutrons, the building blocks of atomic nuclei, with the protons dominating by about a seven-to-one margin. As more seconds ticked by, the Universe continued to expand and cool.

At 3 minutes of age, conditions in the rapidly evolving Universe were ripe for the formation of stable atomic nuclei—varied combinations of protons and neutrons locked together by nuclear forces. For the first time in the (admittedly short) history of the Cosmos, temperatures had "cooled" sufficiently, to a mere 100 billion degrees; the change was enough for nuclei, once formed, to remain intact. Isolated protons—the nuclei of simple hydrogen atoms—would continue to dominate the mix, as hydrogen still does today. Yet hydrogen was not alone. For the next 17 minutes, free neutrons madly combined with whatever protons they could find to form the heavy hydrogen isotope called "deuterium." Most of that deuterium then fused by pairs into a common variety of helium with 2 protons and 2 neutrons, known as helium-4. By the time the Universe was approximately 20 minutes old, it had cooled too much to promote further nuclear fusion reactions. Atomic ratios became more or less fixed. This most simplified version of BBN results in a Universe with about ten hydrogen nuclei to each helium-4 nucleus, and with a little bit of deuterium left over.

That's a useful simplification, but the BBN story isn't quite so straightforward. The nuclear particles (protons and neutrons) also mixed and matched, smashing together in every possible combination to form small but significant amounts of helium-3 (2 protons plus 1 neutron) and lithium-7 (3 protons plus 4 neutrons), as well as some larger unstable nuclei that quickly fragmented. In fact, the ratios of those rare nuclei of helium and lithium observed in today's Universe provide some of the most exacting constraints on calculations of cosmic evolution immediately following the Big Bang. In this version of cosmic origins, BBN produced no stable elements heavier than lithium (Element 3). That statement means no carbon—no Element 6.

This is the funny thing about science. "No carbon" in this context doesn't necessarily mean "zero carbon." "No *significant* carbon" would be a better claim—not enough carbon to affect the subsequent behavior of the stars and galaxies that were to come. Not enough carbon for crystals or atmospheres or oak trees. But our focus on carbon demands that we know about the true birth of Element 6. For us, the emergence of even a single carbon atom holds cosmic significance.

The critical interval between 3 minutes and 20 minutes following the Big Bang was unimaginably violent and intense—a frenetic time of unbridled nuclear interactions and exchanges and consequent atomic novelty. Almost all of the collisions among protons and neutrons made deuterium or helium, but the tiniest fraction of nuclear reactions, especially those between larger nuclear fragments occurring toward the cooler end of that 17-minute interval, produced more complicated combinations of protons and neutrons, including some elements heavier than lithium.

Calculations published in 2007 by Italian astrophysicist Fabio Iocco and his colleagues incorporated more than 100 plausible nuclear reaction pathways that had been ignored in previous efforts as too improbable (not to mention too expensive in oversubscribed supercomputer time) to warrant the requisite lengthy calculations.[2] Iocco's conclusion was this: Yes, those reactions were improbable, but not impossible. Carbon, nitrogen, and oxygen—Elements 6, 7, and 8—

all formed. The amounts were too small to have much effect on the subsequent evolution of the Universe. But form they did. Iocco's calculations suggest that roughly one nucleus of carbon-12 appeared for every 4,500,000,000,000,000,000 (four and a half quintillion) hydrogen nuclei. That seemingly inconsequential fraction was small enough for Iocco and his colleagues to conclude that the earliest stars evolved in a "metal-free environment" ("metal" to an astrophysicist means any element heavier than helium). So, once again, scientists claimed that the Big Bang produced essentially no carbon.

But wait a second. A rough calculation suggests that the Universe emerged from BBN with at least 10^{80} (a numeral one followed by eighty zeros) hydrogen atoms—a staggeringly huge number. At the same time, only one carbon atom formed for every few quintillion hydrogen atoms—a tiny fraction. But a tiny fraction of a huge number can still be a *very big number*. Simple division suggests that the Big Bang produced more than 10^{64} carbon atoms! That total represents only a minuscule fraction of the Universe's mass, and it's only a trillionth of the total carbon atoms found in the Universe today, but it's still a lot of primordial carbon atoms.

Where are those 10^{64} carbon atoms today? Some were surely captured in prior generations of stars, subjected to cycles of nuclear fusion reactions, and thus modified to other, heavier elements. Other Big Bang carbon atoms have been dispersed, scattered far and wide throughout today's Universe in cosmic dust and gas. But vast numbers of those very first atoms of carbon have become intermingled with our modern world, indistinguishable from atoms formed in much later events. Your body contains more than 10^{24} atoms of carbon—100 trillion trillion atoms of Element 6. It follows inevitably that trillions of those atoms must be the very same carbon nuclei that formed so long ago in the throes of BBN—atoms inseparable from the more recent hoard of carbon forged in stars. And the same is true of your essential oxygen atoms and your nitrogen atoms, not to mention all that primordial hydrogen—all elements essential to life.

The startling conclusion is that countless carbon atoms in your body formed not in the stars, as we have long been led to believe, but in the Big Bang, all the way back 13.8 billion years to the beginning of time. Carl Sagan famously observed, "We are made of starstuff."[3] But thanks to BBN carbon, we are all made of "Big Bang stuff" as well.

Star Stuff

Earth and life demand far more carbon—a trillion times more—than could have arisen from the primordial crucible of the Big Bang. To find that great storehouse of Element 6, we must look to the luminous heavens, for almost all carbon atoms were born deep inside stars.

The role of stars in the carbon story began to emerge more than a century ago through the discoveries of a remarkable group of women scientists at Harvard University. Astronomy in the 1880s faced a new challenge: processing an explosion of data on the nature of stars. Previous astronomers had employed the world's best telescopes to record the position and brightness of more than 200,000 stars, but with little accompanying data on their varied physical and chemical characteristics. By the last quarter of the nineteenth century, new methods were enabling astronomers to gain insights by attaching sensitive spectrometers and cameras to powerful telescopes. The resulting glass photographic plates transformed the sky's familiar visual array of thousands of point-like stars into mosaics of stellar spectra. Just as a glass prism spreads out a focused beam of white light into a rainbow band of colors, each star appeared on the photographs as a tiny elongated rectangle with a barcode-like sequence of vertical lines—each pattern representing a rainbow of spectral colors from red to violet.

Such stellar spectra reveal much about a star. When heated to the high temperature of a star's surface, typically between 4,000 and 60,000 degrees Fahrenheit, each chemical element emits its own characteristic pattern of bright lines of different colors to produce a kind of atomic fingerprint. Each line results when an atom's electrons jump from a higher to a lower energy level—a "quantum jump" accompanied by a tiny flash of light at a fixed color. A prominent, closely spaced pair of orange lines distinguishes sodium. Hydrogen, by contrast, has one strong red line, another in the green, and eight weaker lines at the blue-violet end of the spectrum. And carbon's spectrum features more than twenty lines, with the strong bands distributed across all colors. Each stellar spectrum is a complex superposition of the characteristic lines of dozens of chemical elements.

Armed with their new spectroscopic tools, astronomers produced thousands of glass plates, each with hundreds of stars to be analyzed. Each star spectrum had to be examined and interpreted by eye. It was exacting, tedious work. The spectra piled up far faster than anyone could process them.

Thanks to the pioneering studies of physician and amateur astronomer Henry Draper, who made his first photograph of a star's spectrum in 1872, the Harvard College Observatory emerged as one of the most prolific centers for producing glass plates. Draper made more than 100 glass-plate images with stellar spectra, but he died in 1882 as his work was just beginning to gain momentum. Draper's friend, Harvard astronomy professor Edward Charles Pickering, took over the project in 1885. A year later, Draper's wealthy widow, Mary Anne Palmer Draper, began to fund Pickering's research and underwrite publication of the ever-expanding *Draper Catalogue of Stellar Spectra*.

Like most scientific fields in the 1880s, astronomy was an almost exclusively male pursuit. Indeed, for most of the twentieth century in most observatories, women were not allowed to work with men in the supposedly seductive nighttime environment. Men also dominated the job of analyzing photographic plates, though Pickering was repeatedly frustrated by their sloppy performance. "My Scottish maid could do better," he complained more than once.[4]

Fortunately for Pickering, his "Scottish maid" was Williamina Fleming, a teacher who had immigrated with her husband and child to the United States from Dundee, Scotland, at the age of twenty-one. Abandoned by her husband soon thereafter, she resorted to working as a maid for the Pickering household. In 1881, Pickering offered her a job at the observatory, teaching the twenty-four-year-old Fleming to read stellar spectra. His action opened the door to women in the field, though his motive may not have been entirely altruistic. Her salary of 25 cents per hour was significantly less than that of the men she replaced.

Fleming excelled not only at interpreting spectra, but also at seeing patterns among the thousands of stars. She quickly learned to detect subtle differences in the positions and intensities of the varied spectral lines and proposed a classification system, giving each star a letter designation from A to Q based primarily on the strength of characteristic hydrogen spectral lines. She also spotted hundreds of previously unknown astronomical objects, including the famous Horsehead Nebula and dozens of other "nebulas"—vast expanses of dust and gas now known to be rich in carbon-bearing molecules. Fleming also paved the way for more than a dozen other female colleagues at the Harvard College Observatory, who came to be known collectively as the "Harvard computers."[5]

The Harvard Classification of Stars

As new spectral data on thousands of stars poured in, the field of astronomy was poised to profoundly alter our understanding of the origin and distribution of carbon in the Universe. A first critical step was a more nuanced picture of the different kinds of stars—an advance spearheaded by astronomer Annie Jump Cannon.

Annie Cannon was born in 1863 in Dover, Delaware. Her father, Wilson Cannon, was a Delaware state senator and shipbuilder. Her mother, Mary Jump Cannon, loved the night sky. Guided by an old dog-eared astronomy text, mother and daughter together identified stars and constellations. Annie was encouraged to study science at Wellesley College, where Sarah Frances Whiting, the college's first professor of physics, became her mentor. Annie Cannon graduated valedictorian with a degree in physics in 1884 at the age of twenty.

After a decade honing her skills as a photographer and writer, Cannon returned to science, joining Pickering and the Harvard computers in 1896. She rapidly became the master at recognizing different types of star spectra, eventually documenting star types at an astounding rate of 200 per hour. "Miss Cannon is the only person in the world—man or woman—who can do this work so quickly," Pickering marveled.[6] Over a four-decade career, Cannon's lifetime total of 350,000 stars analyzed by eye far surpassed any of her peers' totals.

Cannon's skill at pattern recognition enabled her to see trends that others had missed. Her immersion in stellar spectra provided the experience and insights for a new, revised system of star classification. Focusing on bright stars in the Southern Hemisphere, she devised a grouping based on the relative strength of key spectral lines—information that was directly related to the star's surface temperature. The resulting Harvard classification system divides stars into seven major classes, each with a letter that matches star types in Williamina Fleming's earlier scheme. The resulting sequence from the hottest to the coolest stars eliminated most of Fleming's letters and scrambled the rest: O, B, A, F, G, K, M—remembered by generations of astronomy students with the mnemonic "Oh Be A Fine Girl, Kiss Me."

Cannon was widely lauded for her discoveries in the years prior to her death in 1941, winning medals, fellowships, and honorary degrees in Europe and North America. In the process, she served as a role model for the next generation of women in science.

Why was Cannon so remarkably prolific and successful? Some historians cite the organizing influence of her mother's lessons in home economics. Others point to Cannon's near total deafness, perhaps a consequence of scarlet fever, which may have limited her interest in socializing. But many women of her time dealt with disabilities. And many more learned home economics. I think a more fundamental factor explains Cannon's success: she was brilliant, dedicated, and passionate about astronomy, to be sure, but unlike almost any of her contemporaries, *she was given a chance.* For centuries, the untold narrative of science has been the story of opportunity denied—of unnamed potential Einsteins and Newtons, brilliant minds barred from their birthright, who never discovered their passion because of gender or race. The great tragedy for us all lies in the countless lives unfulfilled, the breakthroughs undiscovered.

Carbon in Stars

Annie Jump Cannon's stellar classification set the stage for discovering the role of stars in carbon formation. The Harvard spectral classification reflects a star's surface temperature, which ranges from relatively cool "red" stars to superhot "blue" stars. Astronomers of the time also knew that spectral lines reveal information about the relative abundances of different chemical elements, but the problem of converting line intensities to chemical composition was daunting and unsolved.

Temperature confuses the matter. Every atom consists of negatively charged electrons in shells that surround the atom's positively charged nucleus. Electrons jumping between these shells produce the characteristic spectral lines that were captured on the Harvard observatory's glass plates. At the high temperatures of stars, however, violent collisions between atoms easily strip off the outermost electrons—the atoms become ionized—thus reducing the intensities of certain lines. Hydrogen and helium, the first and second elements of the periodic table, are extreme cases. Most hydrogen atoms lose their lone electron to become isolated protons; most helium atoms lose their two electrons to become "alpha particles" with two protons and two neutrons. Without electrons, no electron jumps are possible, so the spectral lines of hydrogen and helium ions are much weaker than lines of many other elements.

Cecilia Helena Payne deciphered the complex relationship between a star's spectrum and its composition in a 1925 study described by a group of her peers as "the most brilliant Ph.D. thesis ever written in astronomy."[7] Payne was born in 1900 in Wendover, England, into a family with distinguished academic credentials. Raised from the age of four by her widowed mother, she was encouraged to pursue science. She attended Cambridge University on a scholarship to Newnham College, excelling in biology, chemistry, and physics. At the time, because only men were allowed to attain a Cambridge degree, Payne was denied an opportunity to advance within the British system. Consequently, she left England for the Harvard College Observatory, where in 1925 she became the first woman to earn a PhD in astronomy.

The success of Payne's thesis relied on application of the emerging theory of "ionization"—the temperature-dependent processes by which atoms lose electrons in stars. She realized that while the relative abundances of many key elements—oxygen, silicon, and carbon among them—could be accurately determined from strengths of key spectral lines, the amounts of hydrogen and helium were vastly underestimated, perhaps by a factor of a million in the case of hydrogen. She came to the startling conclusion that hydrogen and helium are by far the most abundant elements in the Universe—in many cases accounting for more than 98 percent of the total mass of a star. This result seemed so improbable to her colleagues, who had long assumed that Earth's composition accurately mimicked that of the Sun, that her findings were initially rejected. Payne was persuaded by senior colleagues to call her own conclusions "spurious" in her first publication, but vindication came soon thereafter, as others replicated her novel methods.

Payne's discoveries pointed the way to a deeper understanding of the cosmic origins and abundance of carbon—an element that accounts for almost one in every four atoms that is not hydrogen or helium. Yet, one fundamental mystery remained. How do stars manufacture that immense quantity of Element 6?

Helium Burning

Most stars are giant hydrogen-rich spheres. Our Sun provides a nearby case in point. The Sun transforms hydrogen into helium for a living—a reliable nuclear fusion process called "hydrogen burning" that has caused our star's brilliance to vary only modestly over the past 4.5 billion years. Ninety percent of stars in the night sky are engaged in this process: helium is manufactured at immense temperatures and pressures deep inside stars, where protons (the nuclei of hydrogen) collide and fuse, creating larger nuclei from smaller bits and pieces. By all accounts, the Sun will be a stable hydrogen-burning star for another few billion years. Only then, when hydrogen in the Sun's core has mostly fused to helium, will a new, more energetic phase of "helium burning"—the process that makes carbon—commence.

English astronomer Sir Fred Hoyle first described the nuclear fusion reactions by which helium transforms to carbon in stars in 1954, while he was a lecturer at Cambridge University's St. John's College.[8] Hoyle enjoyed a career of remarkable diversity. Trained in mathematics at Cambridge, he joined Britain's war effort in 1940 at the age of twenty-five, to work on radar research. His investigations took him to the United States, where he first learned about nuclear reactions from studies associated with the Manhattan Project. Hoyle spent the decade following the war back at Cambridge, immersed in thinking about nuclear processes in stars.

By the 1950s, the basic concept of "nucleosynthesis"—that extreme temperatures and pressures inside stars drive the nuclear fusion reactions that create new elements—was well understood. Hoyle recognized that the natural abundances of elements reflect stepwise stellar processes that fuse smaller nuclear building blocks into larger nuclei. Some elements are common (for example, iron and oxygen) while others are rare (for example, beryllium and boron) because certain combinations of protons and neutrons are easier to make than others. Of special importance are "resonances" that promote the addition of one neutron, one proton, or one alpha particle (a helium-4 nucleus with two protons and two neutrons) at a time. Most new nuclei form by the incremental addition of one of these small nuclear building blocks to existing nuclei.

Carbon was an anomaly. According to calculations at the time, no simple pathway leads to carbon synthesis in stars, so Element 6 should be quite rare. But measurements of carbon concentrations in stars by Cecilia Payne and others suggested that carbon is the fourth-most-abundant element in the Universe. To explain this discrepancy, Hoyle posited a clever mechanism called the "triple-alpha process."[9] Hoyle knew that older stars concentrate helium-4 (that is, alpha particles) in their cores. Two colliding alpha particles easily fuse to form beryllium-8, the nucleus with four protons and four neutrons. All you need to do is add one more alpha particle to transform beryllium-8 to carbon-12. But there's a catch: beryllium-8 is extremely unstable and breaks apart to smaller bits in less than a quadrillionth of a second. Thus, the idea that carbon-12 would form from the addition of a third alpha particle to fragile beryllium-8 seems highly unlikely.

Hoyle's breakthrough was to recognize a coincidence in nature. The carbon-12 nucleus has a specific, previously overlooked resonance at an energy close to 7.68 million electron volts—exactly the value needed by beryllium-8 to snatch up an alpha particle even faster than it decays. Hoyle estimated that the rate of carbon-12 production by this triple-alpha process could be enhanced by a factor of approximately 1 billion. Experimental physicists were skeptical; carbon had been well studied and no such resonance had been reported. Nevertheless, Hoyle convinced researchers at the California Institute of Technology to look for this "Hoyle state," which they confirmed soon thereafter. Hoyle's prediction resolved the carbon abundance discrepancy and, in the process, rocketed him to international prominence in the burgeoning field of astrophysics.

Hoyle gained fame and honors from his explication of stellar nucleosynthesis, but his career was not without controversy. An outspoken critic of prevailing cosmological thinking, he coined the phrase "Big Bang," perhaps as a pejorative term that ultimately stuck. He much preferred the concept of a steady-state Universe, one not reliant on a Genesis-like "moment of creation." Hoyle also invoked "panspermia"—the highly speculative concept that life on Earth was seeded from space. In Hoyle's widely derided version of panspermia, comet-borne viruses started life, and what's more, they still occasionally cause global viral epidemics. And he strongly supported the idea that petroleum and natural gas arise from nonbiological processes deep in Earth's mantle—a controversial hypothesis that Deep Carbon Observatory scientists are now revisiting. When asked about his propensity to adopt contrarian positions, Hoyle replied, "It is better to be interesting and wrong than boring and right."[10]

Dispersing the Carbon

Long, long ago, more than 13 billion years ago, within a few million years of the dawn of creation, the first stars burned brightly in a Cosmos empty of rocky planets and devoid of life.[11] Primordial stars emerged when gravity coerced vast, swirling clouds of hydrogen and helium—themselves the atomic offspring of the Big Bang—into larger and larger glowing spheres.

Stars are the engines of chemical evolution. Subjected to the unimaginable heat and pressure of stellar interiors, hydrogen fused into helium, while triplets of helium nuclei fused into carbon—a slow process, to be sure, but stars have a lot of time. And so, carbon gradually increased in concentration, ultimately to become the fourth-most-abundant element in the Universe, with almost 5 carbon atoms for every 1,000 hydrogen atoms.

For the first few millions of years in cosmic history, the bulk of this ever-expanding inventory of stellar carbon remained locked away deep inside stars. Some carbon nuclei became nuclear fuel, fusing with more helium to forge ever-more-massive elements: oxygen, the giver of animal life; silicon, the builder of rocky planets; iron, the foundation of industry. After millions of years, as turbulent stellar convection brought these deep atomic products to each star's luminous surface, a few carbon atoms escaped in energetic stellar winds, propelled outward by interactions with the stars' intense magnetic fields. Those star-formed atoms, flung into deep space, marked the true beginning of the "carboning" of the Cosmos.

The most profligate seeding of space with carbon occurs when massive stars die—violent processes that liberate immense quantities of matter.[12] In the case of supernovas, big stars literally disintegrate, blasting apart in space. But how can a star explode? The answer lies in the unceasing tension between immense gravity pulling a star's mass inward and energetic nuclear reactions pushing that mass outward.

Consider the future fate of our Sun, which will burn hydrogen to make helium for another 4 billion years or so. Gradually, hydrogen in the Sun's superheated core will be consumed as helium concentrations rise. That's when helium burning will take over. For perhaps a half-billion years, nuclear forces associated with helium burning deep inside the Sun will take the upper hand; the outward nuclear push will dominate gravity's inward pull. That transition will not be a pleasant time to be an Earthling. The Sun will swell to more than 100 times its present diameter, a "red giant" star expanding past the orbit of poor engulfed Mercury, past the orbit of doomed Venus, eventually ballooning close enough to Earth's orbit to fill the daytime sky. As the Sun's ruddy surface approaches Earth, our home will become a desiccated, lifeless cinder.

For the modestly sized Sun, carbon is the nuclear endgame. As helium stockpiles are consumed and nuclear reactions die away, gravity will win the 10-billion-year war. The Sun will collapse into a "white dwarf"—a carbon-rich star about the size of Earth, less than a hundredth of the Sun's current diameter. Slowly cooling and shrinking, most of its hoard of newly minted carbon will be locked away forever, "like a diamond in the sky."

Stars larger than the Sun avoid this fate because their internal pressures and temperatures are sufficient for some carbon-12 to fuse with alpha particles into heavier elements—oxygen-16, neon-20, magnesium-24, and more. A cascade of nuclear reactions ensues, each transformation adding energy to the star, each enriching the star in new chemical elements, and each pushing outward against the inexorable force of gravity. Faster and faster the reactions pile one upon the next, the final stages occurring in seconds, until a star produces iron-56. For elements less massive than iron, each new nucleus is more stable than the last. Each nuclear reaction releases energy and keeps the star blazing, like adding fuel to a roaring fire. But iron-56 is the ultimate nuclear ash. Whatever you do to an iron-56 nucleus—add or subtract a proton, add or subtract a neutron—the reaction with iron consumes energy. When a star's core converts to iron, the outward push of nuclear reactions turns off almost instantaneously, and gravity just as swiftly takes over.

The initial effect of this stellar "off" switch is a devastating implosion with all of the star's mass—all of the remaining hydrogen and helium and carbon and everything else—pulled inward and accelerating to a significant fraction of light speed before everything smushes together. Under those chaotic conditions, with temperatures and pressures skyrocketing to values not seen since the Big Bang, atomic nuclei undergo shocking impacts and violent mergers, shuffling their protons and neutrons to generate heavier and heavier combinations—ultimately generating more than half of the elements in the periodic table. What we observe as a supernova "explosion" is actually the cataclysmic rebounding of all that mass—a jumbled mix containing a host of new elements, flung outward into space as the star disintegrates.

Additional chemical novelty, including most of the heaviest elements of the periodic table, emerges from the surprising aftermaths of supernovas. In processes that are only now coming fully into focus, gravity captures a fraction of the remnants of every supernova to produce strange, dense starlike objects. If those leftovers exceed about three times the mass of our Sun, then a "black hole" results—an object so massive that it collapses into a point from which nothing, not even light, can escape.

If the supernova remains have a mass that is equivalent to only one or two times the mass of the Sun, the resulting gravitational collapse leads to a different beast: a "neutron star," in which protons and electrons crush together to form an ultradense mass of neutrons. A neutron star with a mass twice that of our Sun collapses to an object with a diameter of only a few miles. Given the wide dispersal of atomic bits and pieces following a supernova, it's not uncommon for two neutron stars to form in the aftermath of the same explosive event. The resulting unstable binary star configuration ultimately leads to yet another cosmic catastrophe—an event dubbed a "kilonova"—as two neutron stars collide. The resultant merging of nuclear particles is so energetic that the entire periodic table of the elements emerges from the chaos.

The implications are staggering. The ultimate source of chemical elements heavier than iron—precious gold and platinum, utilitarian copper and zinc, poisonous arsenic and mercury, high-tech bismuth and gadolinium—arises from such cosmic catastrophes. Every atom of those elements found here on Earth arrived via the disruption of massive stars. Tungsten abrasives, molybdenum alloys, germanium semiconductors, samarium magnets, zirconium gemstones, nickel-cadmium batteries, strontium phosphors—all are available to us courtesy of ancient, exploding stars.

Only after that first generation of supernovas seeded the Universe with chemical novelty could rocky planets like Earth arise and encircle the next population of carbon-producing stars. Many of those stars exploded to provide even more carbon and other heavy elements to enrich the formation of more planets and future generations of increasingly "metal-rich" stars. This epic, violent cycle of element creation and dispersal continues to this day throughout the Universe.

Our own solar system, the consequence of many preceding cycles of stars extending backward in time more than 13 billion years, is thus abundantly endowed with carbon, the element of crystals.

EXPOSITION—Earth Emerges and Evolves

A TOMS MIX AND MINGLE, creating crystals of exquisite beauty and remarkable variety. The solid crust, mantle, and core of Earth host vast quantities of carbon-bearing compounds: diamond, graphite, and more than 400 other crystalline, carbon-bearing minerals that comprise the dominant repositories of carbon in Earth. These rich and varied species tell vivid, revealing stories of Earth's sweeping 4.5-billion-year evolution, while their modern synthetic analogs display astonishing diversity and play indispensable roles in today's technological world.

The First Crystals in the Cosmos

Carbon is remarkably gregarious. Every carbon atom is born alone, but carbon atoms cannot abide being alone. They seize any opportunity to bind to as many as four other atoms. Carbon chemistry, the profound consequence of carbon's desperate desire to bond, must have begun early, not long after the dawn of creation. Surrounded by hydrogen, most primordial carbon atoms quickly seized four companions to become molecules of CH_4—methane, or "natural gas."

Carbon chemistry became much more interesting when stars began to explode, littering the heavens with chemical novelty. Prominent among the new chemical elements to emerge was oxygen, a reactive atomic species that binds strongly to carbon. Molecules of carbon monoxide (CO) and carbon dioxide (CO_2) quickly appeared on the scene. Other carbon atoms linked themselves to atoms of abundant nitrogen and hydrogen in deadly hydrogen cyanide (HCN), or bonded to common sulfur or phosphorus in dozens of varied molecular species.

All of those small, primordial molecules formed gases that consorted with hydrogen and helium in great cloud-like nebulas, the nurseries of stars.[13] Carbon also enjoyed the option of linking to itself to build chain- or ring- or cage-shaped structures, as a growing molecular inventory of ever-increasing geometrical complexity ensued. And from time to time, in the most concentrated carbon-rich swirls of a star's expanding atmosphere, each atom of carbon bonded to four more of its own kind in a growing regular array. The result was a tiny crystal of diamond.

Diamond is carbon frozen into crystalline perfection. How can one not love this precious stone? Superlatives abound: hardest, highest thermal conductivity, most brilliant, greatest shear strength, most valued. For centuries, diamond has captured the imagination of consumers and scientists alike. Large, flawless diamond crystals are not simply rare and beautiful treasures—the coveted symbols of love and power. Diamonds are also scientific treasures. They reveal our planet's enigmatic deep interior and preserve our planet's mysterious deep past. They are literally time capsules of Earth's hidden heart and, looking much, much further back in time, they were the very first crystals in the Cosmos.[14]

Here's how it happened. At the high surface temperatures of a carbon-rich star, the vibrations of atoms are much too vigorous and wildly erratic for any pair of carbon atoms to settle down into a stable chemical bond. That situation changes when such a star explodes, releasing a massive, ballooning cloud of gaseous atoms. As temperatures within that expanding gas envelope drop below about 8,000 degrees Fahrenheit, carbon atoms eager for companions slow down sufficiently to bond to four others in tiny pyramids less than a billionth of an inch across. Each of those pyramidal carbon atoms wants four neighbors, so each adds three more carbon atoms, and the new neighbors add even more in precise geometrical alignment. And so a diamond crystal grows.

Microscopic diamond crystallites beyond number have formed in space in this way for billions of years. They formed long before rocky planets, and they continue to form today around most energetic stars in the Universe, crystallizing at the diffuse interface between the incandescent stellar surface and the cold vacuum of space.

On the Remarkable Diversity of Earth's Carbon Minerals

In spite of the ubiquity of space-borne microscopic diamond dust, diamond is not the favored form of carbon throughout most of the Cosmos. At the extreme temperatures surrounding stars, diamond crystallized first because diamond is the only solid that can condense and grow above 8,000 degrees Fahrenheit. Every other crystal melts or vaporizes under such white-hot conditions. But at lower temperatures and pressures, another, more prosaic crystalline form of carbon takes over. The atoms of diamond are too densely packed, too crowded together for comfort. Microdiamonds form readily enough from the cooling gas of a star, but once temperatures drop below about 7,200 degrees Fahrenheit, graphite, the familiar soft black mineral of "lead" pencils and "dry" lubricants, forms instead.

Graphite and diamond are a study in contrasts.[15] Diamond is hard and tough—a consequence of its three-dimensional girder-like atomic framework. In the elegant graphite structure, each carbon atom bonds to three rather than four neighbors in a miniature flat triangle. The result of this less crowded atomic architecture is a layered structure with perfectly planar carbon sheets stacked one on top of the next, like pieces of paper in a ream. These loosely bonded sheets of carbon atoms easily transfer from your pencil to paper, and they slide across one another to lubricate your locks and bearings. Soft, black graphite cannot serve as a gemstone, but its value to society is no less than diamond.

Diamond was the first and graphite, we suspect, was the second crystalline substance in the Cosmos. In spite of their contrasting properties, both minerals are pure carbon, and both initially formed from the residues of stellar violence. Yet the true diversification of carbon-bearing crystals—an explosion of new forms—had to await the formation of the rocky planets, the engines of carbon mineral diversification.

The formation of planets is an ancient, violent process. Vast nebulas, the birthplaces of stars and planets, are tenuous clouds of cosmic dust and gas spanning light-years of space. When disturbed by the passing gravitational wake of a rogue star or the shock wave of a supernova, a small region of the nebula may start to collapse, with gravity pulling the swirling mass inward to rotate faster and faster like a skater in a tight spin. Most of the mass falls toward the center to form a star like the Sun; the leftovers concentrate in a few orbiting planets. In our solar system, the young Sun sustained an intense, hot solar wind that swept most of the remaining dust and gas far out to the orbit of Jupiter and beyond—the distant realm of the "gas giant" planets. The rocky leftovers made the inner planets: Mercury, Venus, Earth, and Mars.

Planets start small, as cosmic fluff balls of dust—microscopic particles loosely held together by static cling. Bursts of solar energy or flashes of nebular lightning melted those clumps into small droplets no larger than BBs. These "chondrules" collected into larger and larger masses—the size of a basketball, then a blimp, then a small mountain.[16] Gravity pulled countless orbiting rocks into larger and larger planetesimals, which merged in increasingly energetic collisions. Fragments of these earliest periods of solar system assembly still fall to Earth as chondrite meteorites—the oldest objects you can hold in your hand. They're not rare; you can buy one on eBay for a few dollars.

As planetesimals grew larger, to 100 miles or more in diameter, their inner heat melted, refined, and separated the raw materials. Dense metals like iron and nickel sank to form planetesimal cores. Less dense accumulations of gemmy olivine and pyroxene crystals mantled the growing worlds. Hot water circulating through cracks and fissures altered the rocky mix, while destructive impacts of big space rocks created new, dense "shock" minerals. By the late stage of this process, a few large protoplanets—Earth among them—dominated the emerging solar system, sweeping up most of the remaining rocky debris like immense vacuum cleaners. A final epic collision between Earth and its smaller companion, the protoplanet Theia, resulted in the annihilation of the latter and the formation of Earth's luminous Moon.

As the Moon coalesced in the sky, the molten, injured Earth quickly healed and cooled, forming a thin, brittle crust, a thick mantle, and the inaccessible metal core. Deeply circulating superhot water and steam selected and concentrated chemical elements, transporting them to the young planet's cooler surface, where they formed a growing inventory of mineral novelties—many carbon minerals among them.

Primordial Earth, pummeled by space rocks bearing both diamond and graphite, had barely begun its flamboyant exploration of the sixth element. Gradually, as Earth evolved, so did its astonishing carbon mineralogy—hundreds of crystalline forms, each a unique combination of chemical composition and crystal structure, each incorporating carbon in varying combinations with other companion chemical elements. Each of these astoundingly diverse minerals bears witness to our dynamic, evolving world.

—

Today, carbon-bearing minerals abound.[17] From the immense limestone peaks of the Canadian Rockies to the vast coral platforms of the Great Barrier Reef, from the White Cliffs of Dover to the untold accumulations of tiny shells on the ocean floor, minerals hold the crust's greatest reservoirs of Element 6. More than 400 different kinds of known mineral "species" incorporate carbon. What's more, our recent research suggests that many more prizes remain to be recognized and described; more than 150 new species of carbon-bearing crystals lie hidden around us embedded in rocky outcrops, lining the sides of superheated volcanic vents, growing along the shores of evaporating lakes, and littering abandoned mine dumps. These rare crystal forms await discovery.

The sheer diversity of carbon minerals amazes. Their colors span the rainbow—flaming red, intense orange, vibrant yellow, stunning green, startling blue, and rich violet. They come in every hue and shade: whites, grays, tans, and blacks—some perfectly transparent, others diaphanous, translucent, or opaque. Their lusters, too, are richly varied—metallic, flat, sparkling, resinous, waxy, milky, or iridescent. The forms of carbon minerals are no less diverse: elegant, faceted crystals in cubes and octahedra, sprays of slender needles and clusters of flat plates, gently tapered or sharply pointed; shapeless lumps, rough crusts, sensual rounded forms, and irregular jagged masses—all in sizes from microscopic to larger than a beach ball.

In Earth's dynamic crust, most carbon atoms link to three oxygen atoms, forming a tiny, flat triangle—a four-atom cluster known as a "carbonate group." These atomic building blocks form the richly varied carbonate minerals, perhaps most familiar to us as the sturdy shells of snails and clams, as dietary calcium supplements, as marble countertops, and in lustrous pink rhodochrosite jewelry.

Carbonate minerals, especially sedimentary layers of limestone and dolomite, represent by far the largest repository of carbon in Earth's crust—perhaps 100 million billion tons of carbon.[18] That's more than a thousand times the amount in all the other crustal reservoirs—all the coal and oil, oceans and atmosphere, plants and animals—combined.

It would be hard to imagine modern society without these diverse carbon-bearing minerals and their myriad synthetic analogs. They play central roles in smelting iron, forging steel, fertilizing crops, producing glass, and making cement. They help in the manufacture of products as diverse as detergents, fireworks, ceramics, pharmaceuticals, surgical tools, explosives, jewelry, and baking soda. They reduce the acidity of tap water and remove pollutants from power plants. They serve as abrasives for the most efficient machining tools and lubricants for the most demanding industrial applications. What's more, the flamboyant variety of natural carbon-bearing crystals hints at a wealth of plausible synthetic materials with even greater potential—with engineered properties fine-tuned to our hopes, our needs, and our desires.

An exploration of those varied minerals—their myriad forms and hidden origins—reveals much about carbon and the way our planet cycles and stores the vital element. We have begun to catalog that richness, to probe deeper and deeper into Earth, even to predict what might be missing from our still-incomplete inventories. Carbon mineralogy is a theme with many variations spanning centuries of research and discovery.

To understand that history, we must journey back two centuries, to Scotland at a time when carbon-bearing minerals lay at the heart of a seemingly unresolvable geological controversy.

Carbonate Minerals Reveal Earth's History

Society depends on limestone—a rugged, gray, carbon-rich rock that forms prominent cliffs and jagged mountains across the globe. These abundant ancient sediments were deposited layer by gradual layer, sometimes as accumulations of coral and shells, sometimes as chemical precipitates from ocean or lake waters rich in calcium. Billions of tons of crushed limestone are sold every year as durable foundations for highways, railroads, buildings, and bridges—a geological resource with annual sales exceeding those of diamond, silver, or gold. You may have bought more modest amounts to enhance your walkways and patios.

Blocks of limestone, and their more densely crystallized cousin marble (a form of limestone deeply buried and transformed by Earth's high pressure and temperature), create imposing buildings and monuments; the pyramids at Giza in Egypt and the Lincoln Memorial in Washington, DC, come to mind. Fancy varieties of limestone, often adorned with fossil shells, are commonly employed as "dimension stone" to face buildings, tile floors, and decorate kitchen countertops. You've probably used powdered lime on your garden or lawn to correct acidic soils, and you may have taken calcium as a dietary supplement. Chickens (and chicken farmers) also benefit from powdered limestone supplements, which make eggshells stronger and less likely to crack in transit to your local grocer.

The carbonate minerals of limestone also serve as the starting point for a variety of manufacturing tasks. Foremost among them is the production of lime (chemical name "calcium oxide"), which results when limestone is heated in a kiln to roughly 1,800 degrees Fahrenheit. Lime (not to be confused with the limestone powder you put on your lawn) is incredibly useful. It is the major ingredient of mortar, plaster, and cement, forming hard and durable solids when mixed with water. Lime provides the "white" of whitewash. And for thousands of years lime has been a principal ingredient in the smelting of iron and other metals, providing the flux that chemically separates impurities from the metal. In every industrialized country, historic limekilns, many of them begun hundreds of years ago as mom-and-pop operations, dot the countryside.

—

The manufacturing of lime from limestone, a process familiar to any geologist of the eighteenth century, played a curious role in the history of science. In a real sense, limestone threatened to set back Earth science by decades.

In the mid-1700s, a debate raged among European scholars regarding the relative geological roles of water (championed by the "Neptunists") and heat (the favored mechanism of the "Plutonists") in producing rocks.[19] The Neptunists, some with clear biblical creationist leanings, saw the Great Flood—a catastrophic global event that took place within the 10,000-year time frame of the Bible's literal chronology—as the major agent of geological change. Plutonists viewed heat arising from volcanic activity as an equally important agent of change, though they required much longer time intervals to create the modern landscape.

The seeds of the debate sprouted in continental Europe, where geologists studying water-deposited sediments quite naturally favored water, while those who studied volcanic lavas favored fire. The controversy even made it into the act 4 dialogue of Goethe's influential play *Faust*, with the Plutonist perspective poorly defended by the devil himself. By century's end, the focus of the scientific debate, as well as its eventual resolution, had shifted to the enlightened city of Edinburgh, Scotland, and the transformative field studies of James Hutton.[20]

Hutton was born in Edinburgh in 1726, one of five children of Sarah Balfour and William Hutton, a wealthy merchant who died when James was only three. James's mother emphasized the importance of education, and the young James responded, showing a special aptitude for math and chemistry—interests that would serve him throughout his life. Following advanced studies in the humanities, philosophy, and medicine at universities in Edinburgh, Paris, and Leiden in the Netherlands, Hutton went to London with the hopes of establishing a lucrative medical practice. Unable to secure enough patients, he returned to Edinburgh and went into business for himself. He had previously developed a novel chemical process to extract ammonium chloride, widely used as a fertilizer, from the abundant soot and ash produced by Edinburgh's many furnaces and factories. He put his new method into production, operating a profitable chemical factory in Edinburgh.

With his financial future secured, Hutton turned his attention to a new interest, agricultural chemistry. He had inherited two family farms, and he used those properties to conduct experiments to improve agricultural yields. In the process of working with the varied rocks and soils of Earth's crust, he began to think about geology.

The rocks of Scotland display a rich variety of characteristics—sedimentary rocks and volcanic rocks, some fresh and flat-lying as when they formed, others disrupted and deformed. Hutton was also within a day's ride of metamorphic terrains, glacial deposits, and a host of igneous rocks. Of special interest were the sea cliffs of Siccar Point near Jedburgh, Scotland, where Hutton studied the striking juxtaposition of layers. There he observed, vividly exposed by the eroding action of wind and waves, gently sloping beds of younger red sandstone and pebbly sediments overlying steeply inclined layers of older, darker sandstone. The boundary between these layers is sharp, as if the set of near-vertical layers had been sheared off before the horizontal layers were deposited. How could such a distinctive geometry have arisen?

Hutton realized that every aspect of the Siccar Point cliffs—indeed, every aspect of Scottish geology—could be explained simply as resulting from the gradual natural processes at play around us all the time. On the one hand, new sediments are deposited ever so slowly in layers and are gradually buried, heated, squeezed, and turned to stone—all processes that add to the rock record. On the other hand, older rocks are gradually deformed, uplifted, and eroded away—a net subtraction of rock layers. Siccar Point reveals all of these processes in one telling view: older sediments were laid down in flat layers, buried, and turned to stone. Deep forces compressed the layers, distorting them into tight vertical folds. Uplift eroded away the top surface of those older rocks. Another cycle of burial and sedimentation produced the younger flat-lying red sandstones, after which another episode of uplift exposed the red layers to erosion.

There's nothing particularly exotic or new in Hutton's explanation, save one thing: deep time. Others thought of Earth history in terms of a few thousand years. Hutton was comfortable with hundreds of millions, even billions, of years of uniform, gradual change. He saw in the rocks of Scotland "no vestige of a beginning, no prospect of an end."[21] Hutton's two-volume 1795 work titled *Theory of the Earth*, though presented in a turgid style that may have muted its initial impact, initiated a paradigm shift in science.

In these pursuits, Hutton was greatly influenced by the spirit of empiricism that characterized much of the dynamic Scottish Enlightenment. He benefited from interactions with dozens of intellectuals, both at the Royal Society of Edinburgh and at local clubs frequented by the likes of poet Robert Burns, economist Adam Smith, and philosopher David Hume. But it was Scottish geologist and geophysicist James Hall who became the experimental hero of Hutton's story.[22]

James Hall and the Great Limestone Controversy

Like so many of his contemporary scientists, James Hall was born into a rich, aristocratic family. His studies of geology, chemistry, and natural history at the great universities in Cambridge and Edinburgh were undertaken in the context of wealth and privilege. He traveled widely in Europe, buying scientific books for his library and meeting with French researcher Antoine Lavoisier, one of the founders of modern chemistry. Biographers seldom fail to give Hall his full title—Sir James Hall of Dunglass, 4th Baronet—though his fame rests far more on his scientific discoveries than on any aristocratic lineage or label.

Returning to Edinburgh from his travels, Hall learned about the development of his friend James Hutton's revolutionary ideas first-hand. Hutton's *Theory of the Earth* rested on a variety of geological phenomena, among them the interaction of molten lava with layers of sediments—a scenario requiring the juxtaposition of Neptunist and Plutonist processes. Hutton realized that when volcanoes erupt, molten rock oozes upward through old sediments while tongues of lava penetrate between layers deep underground. Such intrusive events were prominently on display at several Scottish localities, notably Arthur's Seat in Edinburgh's Holyrood Park, a glacially eroded hill that affords a textbook example of such interactions, not to mention a magnificent view of the city to the west.

The challenge to Hutton's theory came from localities where similar molten rocks penetrated limestone. How could limestone survive the heat of molten rock, his Neptunist opponents asked? Everyone knows that superheated limestone must turn to lime, just as in a lime-kiln. Therefore, basalt, granite, and other purported "igneous rocks" cannot have been hot; they must have been precipitated from water at more or less the same time as the limestone. To some scientific bystanders, this challenge seemed an irrefutable, fatal blow to Hutton's theory. Hutton countered that limestone, when subjected to the high confining pressures of burial, must remain unchanged even at high temperature. But how could anyone test these speculations in the laboratory?

In spite of his skepticism of the Plutonist theory, Hall devised an imaginative experimental solution to resolve this conflict. Putting his friend's hypothesis to the test in a series of remarkably original experiments, Hall became the original pioneer of deep carbon research. In

a candid admission, he wrote, "After three years of almost daily war-
fare with Dr. Hutton on the subject of his theory, I began to view his
ideas with less and less repugnance."[23] In one set of experiments, Hall
applied high temperatures to basalt and granite to see how they would
behave. Would they break down like limestone, thus disproving their
volcanic origins? As Hutton had predicted, Hall's rocks melted to red-
hot lava and cooled back to their original state—a key property of any
supposed igneous rock.

In follow-up experiments in 1798, a year after Hutton's death,
Hall's great innovation was to apply pressure to his heated samples.
To do this, he packed sawed-off gun barrels with limestone and clay,
welded the barrels shut, and placed the assemblies into a hot furnace. The expansion of gases released by heat generated high internal
pressures—much, much higher than the pressure at Earth's surface.
Many of Hall's experiments failed when the welds leaked or the metal
split, and at least one trial in which the reactants had been insuffi-
ciently dried ended in a disastrous explosion. "The furnace was blown
to pieces," Hall wrote. "Dr. Kennedy, who happened to be present on
the occasion . . . scarcely lived."[24]

But some of Hall's limestone-packed gun barrels held tight, and he
showed that limestone under pressure can be heated to high temperatures, even above its melting point, without decomposing to lime. Hall
presented his transformational work, "Account of a Series of Experiments, Shewing the Effects of Compression in Modifying the Action
of Heat," at an 1805 meeting of the Royal Society of Edinburgh (of
which he would become president seven years later). James Hall not
only vindicated Hutton's ideas, but he ushered in an age of high-
pressure research—an enterprise that flourishes to this day, shedding
light on Earth's deep carbon cycle.

Earth's Rarest Minerals

Limestone, composed of the ubiquitous calcium carbonate mineral
calcite, is by far the largest repository of carbon in Earth's crust, but

calcite is only one of hundreds of documented carbon-bearing minerals. If we are to understand carbon in Earth, then we must shift our attention from commonplace minerals like calcite to more exotic mineral species, each of which possesses a unique combination of chemical composition and crystal structure. We must focus on some of the rarest crystals on Earth.

If you are to learn nature's secrets, you must become intimate with nature's data. Every scientist has known periods of obsession, of spending days pondering graphs and absorbing tables, filling one's mind with pages of detail. It's a kind of temporary insanity—poring over lists while eating, while pretending to converse with colleagues and family, falling asleep thinking of numbers and waking up thinking of numbers. If you're lucky, if you become open to hidden patterns, if your brain makes the right connections, you may see something new—something that no one has seen before.

I admit, I have known such times. In the summer of 2015, I spent days immersed in the details of all known mineral species—more than 5,000 types. I steeped myself in their complex chemistries and intricate crystal structures, their properties and modes of formation, their varied localities and mineral associations. For days I withdrew from pressing concerns. Colleagues chafed at unanswered emails. Family increasingly mirrored my inattention, my aloofness.

Five thousand species is a lot, but it is possible to survey them all in an obsessive week of focus. In a week, you can get a "feeling" for the sweep and richness of the mineral kingdom. What struck me most forcefully is how few minerals are common. Fewer than 100 species make up almost all of Earth's crust—certainly 99.9 percent of its volume. By contrast, most minerals are exceedingly rare, and carbon minerals are no exception.

Carbon mineralogy embraces a lot more than diamond and graphite. Geologists tabulate more than 400 carbon-bearing mineral species, each with a unique combination of carbon and other chemical elements, each with a distinctive geometrical arrangement of atoms in a regularly repeating crystalline structure. A few of these species are found abundantly on every continent: ubiquitous calcite of limestone cliffs and classroom chalk, aragonite of coral reefs and clamshells, mountain-forming dolomite, and the utilitarian magnesium ore magnesite. Carbon also contributes to beautiful "semiprecious" colored stones—one notch below rubies and emeralds in the scale of desire: delicate pink rhodochrosite, rich green malachite, and deep-blue azurite, my favorite mineral.

But for every common carbon mineral, we find a dozen more obscure species—minerals that most people, including most mineralogists, have never heard of. A host of incredibly rare, microscopic crystals are known from only one or two places in the world. Tiny crystals of purple abelsonite were extracted from drill cores from only one small district of the 50-million-year-old Green River oil shale in Colorado and Utah. Beautiful sky-blue juangodoyite is unique to the Santa Rosa silver mine in Chile's Iquique Province. Crystals of gorgeous, emerald-green widgiemoolthalite (try saying that fast three times!) have been unearthed exclusively from the Mount Edwards Mine in Widgiemooltha, Western Australia. And Earth's entire documented supply of grootfonteinite, found as microscopic grains from the Kombat mine in Namibia, would fit into a thimble with plenty of room to spare.

Why are so many mineral species rare? Why wouldn't Earth's atoms find a few dozen optimal arrangements and stick to those? It's a question that my colleagues and I had never thought to ask. And that's why it's so important to have clever, inquisitive, broadly informed, and passionate friends who don't share your own expertise. Like carbon, the more diverse bonds we form, the richer our potential. It's vital to have colleagues in other fields—thinkers who aren't afraid to ask the truly original questions that the experts in your field never think to ask. For me, Jesse Ausubel is such a friend and colleague.

Jesse calls himself an "industrial ecologist"; he studies the supply and flow of energy through societies. He is known around the world for his expertise and his provocative and informed views on energy policy, but his faculty position at the Rockefeller University in New

York City provides a platform for a much more diverse and creative range of intellectual pursuits. He's an expert on the art and life of Leonardo da Vinci. He has proposed novel and persuasive theories on phenomena from mass extinctions to airline disasters. Jesse is an authority on the diversity and distribution of marine life, and he has expertise in the use of DNA fingerprinting to identify species of plants and animals—a global effort called the "Barcode of Life."[25]

Jesse is also an amazing mentor of young scientists, training them in sophisticated procedures and nurturing them through wildly original projects. In 2011 he advised teenagers Catherine Gamble, Rohan Kirpekar, and Grace Young of Manhattan's Trinity School in a study of the ingredients of tea.[26] It turns out that many tea recipes are secret—key ingredients never revealed. But DNA testing can tease out even the minor components of any tea, from Lipton to the most exotic Asian blends. Employing Barcode of Life techniques, the young sleuths found a surprising range of unlisted additives—parsley, bluegrass, alfalfa, and the common weeds white goosefoot and red bartsia among them.

A year later, in 2012, Jesse advised high schoolers Kate Stoeckle and Louisa Strauss, also students at the Trinity School, who posed as diners at more than a dozen high-end sushi restaurants and fish stores. Unbeknownst to the proprietors, Stoeckle and Strauss took small samples of raw fish back to the lab for DNA fingerprinting. The results were devastating: one in four fish was mislabeled, always as a more expensive delicacy: everyday cod marketed as red snapper; roe from common smelt sold as roe from flying fish; cheap tilapia lovingly presented as pricey "white tuna." When New York's top news media picked up the story, a furor ensued. Dubbed "sushigate," the findings embarrassed prestigious Japanese restaurants and led the Food and Drug Administration to impose new regulations for testing and labeling fish.[27]

Jesse Ausubel certainly understood the social implications of research on tea and sushi, but he saw a bigger story as well: the opportunity to bring science into people's daily lives. Quoted in the *New*

York Times, he said, "Three hundred years ago, science was less professionalized. Perhaps the wheel is turning again where more people can participate."[28]

Jesse had changed my life as the Sloan program officer who facilitated the Deep Carbon Observatory. Fast-forward through years of preliminary grants, workshops, proposals, and team building and we're now nearing the culmination of the Deep Carbon Observatory's efforts. Throughout DCO's evolution, Jesse remained a dynamic, hands-on colleague, fully engaged in the science and its planning.

One memorable interaction occurred in October 2015, after an intense day in the famed volcanic crater Solfatara di Pozzuoli near Naples, Italy. This active zone of venting carbon dioxide and acrid sulfur-rich vapors is the site of complex mineralization. Beautiful red, orange, and yellow crystals condense directly from hot gases rich in sulfur, arsenic, mercury, and other noxious elements. The power of smelly volcanic fumes to make crystals was remarkable and utterly new to me; I had never seen such a mineralogical spectacle.

That evening, riding through Rome in a taxi, Jesse and I discussed the rich variety of minerals and their skewed distribution with so many rare species. That's when he asked the question most geologists wouldn't ask: Why are so many minerals rare? Immediately it struck us that if we are ever to understand all the forms of carbon, we had better get a grip on the numerous uncommon carbon-bearing crystals. Why are there hundreds of rare carbon minerals, each with a unique combination of chemistry and structure? The conversation flowed from there, building on Jesse's knowledge of exotic marine life and my expertise in mineralogy. By the end of the taxi ride, an idea for a paper had taken form.[29]

Jesse and I realized that minerals might be rare for any of four distinct reasons. Thousands of mineral species are rare because they incorporate one or more rare chemical elements that must be selected and concentrated before a mineral can form. That's why relatively few minerals feature cadmium, iodine, rhenium, or ruthenium—all elements present at concentrations of less than one in a billion atoms in Earth's crust. Odd, unlikely combinations of elements also lead to great rarities. Beryllium plus antimony occur together in only one mineral, welshite. Named for Bill Welsh, an avid mineral collector (and, to my great good fortune, my eighth-grade science teacher), welshite is found exclusively in the historic Långban mining district in Sweden. Similarly, vanadium and molybdenum are coupled in only one mineral from one locality: hereroite from the Kombat lead-zinc mine in Grootfontein, Namibia.

A second group of mineral oddities employs abundant elements, but rarity arises from the exacting conditions necessary for their formation. Calcium, silicon, and oxygen are among the most common elements on Earth, but the mineral hatrurite, with a 3-to-1-to-5 ratio of those elements, has been found just once, in Israel's Hatrurim Formation. Hatrurite crystallizes only in a narrow range of composition and at unusually high temperatures (above 2,100 degrees Fahrenheit). Change the conditions even slightly, especially if the common element aluminum is present, and other minerals form instead.

Quite a few minerals are rare because they are ephemeral: once formed, they quickly disappear. The manganese chloride scacchite absorbs water from the air and crumbles away on a humid day. My namesake mineral, hazenite, found only at Mono Lake in California, dissolves every time it rains. (The pretty little crystals are a form of microbial poop; as a colleague says, "Hazenite happens.")[30] Other rare species dehydrate in air, decompose in sunlight, or simply evaporate. Some of these minerals must occur often, at many more localities than are reported, but you have to be in the right place at the right time to find them.

And finally, some minerals are infrequently reported simply because they are much too remote or much too dangerous to collect. Specimens from active volcanoes or deep mines, frozen in Antarctic ice or sequestered far beneath the ocean floor, may be quite common, but they are unlikely to make it into museums, much less the collections of amateur mineral sleuths.

For a mineral to be rare, as is the case for more than half of all species, it has to possess at least one of these four characteristics: quirky composition, finicky formation conditions, an ephemeral lifetime, or a dangerous environment. In a very few instances, an exceptionally rare mineral displays all four traits. A case in point is fingerite, a mineral named for my longtime Geophysical Lab colleague and mineralogical mentor Larry Finger.[31] Fingerite is rare because (1) it possesses the unusual element combination of copper and vanadium; (2) it requires a precisely 2-to-1 ratio of those elements (if the ratio is 1.5 to 1 or 2.5 to 1, other, equally rare minerals form); (3) it dissolves away every time it rains; and (4) it forms only in superheated vents of scalding steam near volcanic summits. It's little wonder that fingerite is known from only one place on Earth—in pockets of hot gas near the top of western El Salvador's potentially dangerous Izalco Volcano.

The realization that most minerals are rare has profound implications. Rare minerals point to places on Earth where the combination of chemical and physical conditions was unusual, if not unique. Perhaps hot brine rich in dissolved nickel and copper was squeezed into a crack in limestone a half mile below the surface. The result: a few rounded crystal clusters of deep-green glaukosphaerite blossomed. Tiny rosettes of yellow bayleyite and crusts of tan swartzite grow exclusively on the walls of uranium mines, while golden needles of hoelite and transparent blades of kladnoite condense only in the vicinity of coal mine fires.

Our dynamic, living planet plays many similar quirky tricks in many unexpected environments. Odd mineral niches, the result of chemical mixing driven by inner heat, migrating fluids, and the pervasive, irrepressible action of life, lead to Earth's unique mineral "ecology." And so our home differs from any other known world. Earth's shining Moon, swift planet Mercury, even our red neighbor Mars, which was by all accounts once warm and wet like Earth—all pale in their mineral diversity by comparison.

Rare minerals provide a scientific feast for those of us who make a living studying nature's crystalline realm. For one thing, most rare minerals possess previously unknown crystal structures—new geometrical arrangements of atoms that inform the quest for new and useful materials. Rare minerals also incorporate previously untested mixtures of elements. Such novelties also guide efforts to invent new materials. But perhaps most surprising, documenting all the rare minerals turns out to be the key to predicting the myriad minerals not yet identified—the rarities that must occur at or near Earth's surface but have yet to be discovered and described.

Big-Data Mineralogy[32]

The secret to predicting which minerals are missing lies in comprehensive studies of the minerals that are known. The Deep Carbon Observatory needed a complete inventory of the hundreds of carbon minerals—the common and the rare, each a distinctive form of carbon in Earth. We also needed to know their quantities—a list of all their worldwide localities and all of the coexisting minerals at each of those mines, quarries, mountain peaks, and tidal flats.

Big-data mineralogy is the key to the prediction of Earth's undiscovered mineral realm. We must construct databases of the more than 5,000 known mineral species and the millions of localities around the globe where those species are found. Then we must mine those data to find hidden patterns that point us in the right direction for discovery.

It takes a special person to nurture a big database—someone who loves the subject, who has a creative vision, who possesses technical expertise in the development of database software, and, perhaps most important, someone who is willing to put in the countless hours essential to seeing the project through. Robert Downs, professor of mineralogy at the University of Arizona in Tucson, fits the job description.[33] He has devoted two decades to building the world's most comprehensive list of minerals and their propertics.

Bob Downs is not the first person you'd think of when considering this herculean challenge. A mellow Canadian by birth and temperament, he came late to the science game. He excelled in math at the University of British Columbia but was just as happy as a construction worker, building highways in the Northwest Territory, subways in

Vancouver, and railroad lines in British Columbia. For a time, Downs dug for gold at his father's claim on Fifteenmile Creek in the Yukon, and he mined fine mineral specimens at his own mountaintop prospect in southern British Columbia, blasting hard rock with dynamite. "I got the dynamite for free 'cause I knew someone." He adds, "I was an idiot but lucky, so I didn't get killed." Only after a lifetime of adventures did Downs settle down to complete his doctorate in mathematical crystallography at Virginia Tech at the ripe old age of thirty-seven. Three years later, after a stint as postdoc working with me at the Carnegie Institution, he joined the faculty at the University of Arizona in Tucson.

On the surface, Bob is an easygoing man, but dig deeper and there is a strong passion for minerals and their study. Decades ago, he realized that mineralogy, with data on its thousands of species scattered in hundreds of sources, lacked a necessary degree of order and rigor—a systematic listing of all officially approved mineral species, accurate determinations of their crystal structures, and a comprehensive tabulation of their physical and chemical properties. So he began what was to become the world's most comprehensive database of mineral species.

At first it was a quiet, small-scale labor of love—a personal compilation of the best crystal structure data on minerals. As the crystal-structure editor for such leading journals as *American Mineralogist* and *Canadian Mineralogist*, Downs had firsthand access to hundreds of tables of crystallographic data. He also realized that there was no comprehensive list of officially approved mineral species. The International Mineralogical Association, universally known as the IMA, is tasked with vetting submissions for new species from around the world—to verify each unique combination of chemical composition and crystal structure found in nature. But the IMA is a volunteer organization, and for many years its "official" list was a rather informal hodgepodge, neither systematically updated nor published with regularity in any one place. So, Downs began his own list, coordinating with the IMA to bring more order to the field.

Money transformed the effort. Downs's angel was Mike Scott,

the billionaire founding CEO of Apple Computers, who joined Steve Jobs and Steve Wozniak in 1977, when the fledgling company was being run out of a garage. Scott is a passionate collector of magnificent gemstones, with an accumulation of big, perfect, deeply colored treasures—a collection that surpasses that of almost every museum in the world. He wanted to develop rapid and unambiguous methods to identify cut gemstones, so he approached Downs with a deal. Scott would contribute $5 million to add state-of-the-art instruments to Downs's lab and support development of a database of minerals and their properties. Downs would help identify Scott's minerals. One proviso: the database had to be named after Mike Scott's cat, Rruff. Thus was born the RRUFF mineral database.[34] Some of us have noted that calling the contents of this resource "rruff data" might make a less than positive impression. But that was the deal and the RRUFF name has stuck. (Check it out at http://rruff.info/ima.)

The idea at first was to gather lots of data, not necessarily to document every single mineral species. But momentum kept building. Downs hired an army of undergraduates to enter vital mineral data, to measure atomic structures and optical properties, to archive characteristic specimens in the growing University of Arizona mineral collection, and to take magnificent microscopic photographs of characteristic crystals. He hired programmers to streamline the data entry process, to add new data fields, to link to other mineral data resources, and to make the site more user-friendly. He welcomed a succession of graduate students, who ended up building their careers on the collection and use of mineral data.

In the process of building the RRUFF mineral database, Bob Downs has become the indispensable man of mineralogy. His website, updated every few days, has the most complete listing of mineral species in the world. Rruff.info/ima garners almost 100,000 hits every week, as the mineralogical community—a mix of students, faculty, amateur collectors, and museum curators—flocks to his website.

RRUFF is still expanding in content and scope. You can now search for minerals by composition, structure, or mineral group.

Downs and his colleagues recently added a "mineral evolution" feature with almost 200,000 ages of minerals layered onto other data. And new statistical packages and graphical options let users visualize mineral data in clever ways. Downs's mineralogical expertise landed him on Mars as a member of the Mars Science Laboratory team that operates the rover *Curiosity* on the red planet. As a consequence, planetary mineralogy has been added to the database as well.

Now everyone has free and open access to the complete catalog of more than 5,000 IMA-approved minerals on Earth and other worlds, with links to all manner of vital statistics about each species. But knowing the myriad forms of carbon that have been discovered is only the first step in predicting what's missing. We also need data on the world's hundreds of thousands of mineral localities—all of the mines and mountains, quarries and outcrops, caves and cliffsides. Building such a list—though a much, much more extreme challenge than is cataloging the 5,000-plus species—is the only way we could discover the quantities of Earth's carbon-bearing minerals.

Mindat.org

A second hero of the international mineral database effort is Jolyon Ralph, a dedicated, no-nonsense Brit who has built a small empire out of gathering and sharing mineral and gemstone data.[35] Like so many of us in mineralogy, Ralph started collecting as a child. He recalls his very first specimen: crystals of quartz in a pebble from the famed Tintagel coast of Cornwall, England, collected when he was six. It was the start of a lifelong love affair with minerals. His collection has grown, but he still has the small stone forty years later.

His second passion began in 1980 when, at the age of ten, he was chosen to participate in a pilot program to teach British children computer programming. His enthusiasm for coding has continued unabated. He enrolled at the prestigious Royal School of Mines as a geology major, but soon switched to computer science as his chosen profession.

Mindat.org, now the world's most extensive mineral-locality data resource, and the essential complement to Downs's RRUFF mineral species database, began on Christmas Day 1993 as Jolyon Ralph's personal mineral list.[36] At first, it was simply a catalog of his own specimens and collecting localities, but Ralph gradually realized that his database could become something greater. He kept adding data and enhancing the features of Mindat as the idea of creating a site that contains all mineral species from every locality around the world began to take shape. The advent of Windows operating systems and the power of the internet provided new impetus to expand Mindat, which went public on October 10, 2000.

There was no way one person could pull this off. Hundreds of years of mineralogical research, amplified by the enthusiasm of tens of thousands of knowledgeable and passionate collectors, has produced a flood of data on mineral species and their localities. Millions of facts detailing which suites of mineral species are found at which localities lie scattered, many buried in countless books and articles published in scores of languages. An unknown quantity of additional valuable mineral data is "dark"—unpublished scraps of information accumulated on index cards, secreted in file drawers, scribbled in field notebooks, and stored on obsolete computer drives. Jolyon Ralph's mission is to find all those data, to organize them into one seamless internet platform, and to give them to the world.

Many people care about the exact locations where minerals are found. Geologists want to know where to go to learn how minerals form. Collectors want to know where to go to find the best display specimens. And mining companies want to know where to go to make a lot of money. Until relatively recently, there was no consolidated source of mineral locality information. A few mineralogists have produced regional studies; geology libraries are filled with titles like *The Minerals of Arizona* or *The Minerals of the Carpathians*. Other reviews have titles like *Mines of Death Valley* or *Gemstones of the World*. And popular mineral periodicals like *Mineralogical Record* and *Rocks and Minerals* publish lavishly illustrated stories on the world's greatest mineral collecting sites, often with carefully compiled lists of all documented species from a productive mine or famous geographical region. But to obtain a comprehensive overview—to tabulate every known occurrence of a relatively common mineral like azurite or rhodochrosite, for example—you'd need years to examine many thousands of sources, a significant fraction of them obscure documents in foreign languages. You'd also need a legion of eager helpers, and that's just who Jolyon Ralph has recruited.

The Mindat statistics are staggering. Almost 50,000 registered users can contribute mineral photographs, describe mineral localities, or edit mineral data. They have collectively uploaded hundreds of thousands of photos of mineral specimens. Data have been recorded for 300,000 localities worldwide, and the number of individual mineral occurrences has surpassed 1 million. "I never expected it to grow as much as it has," Ralph marvels. "It began to take over my life!"

Ralph, who now works full-time on managing and expanding Mindat, is the first to admit that there's more to do. Coverage from some mineral-rich geographical areas—China, in particular—is spotty. Many localities are poorly described (he's trying to add GPS coordinates to all of them). What's more, there will always be nagging errors and biases in a crowd-sourced resource like Mindat. Collectors sometimes make mistakes in identifying their finds, and they're a lot more likely to report rare species that occur in colorful crystals than the much more common white or gray rock-forming types. Nevertheless, through decades of effort and organization, Jolyon Ralph's Mindat. org has transformed mineralogy and opened startling new avenues for mineralogical research.

Mineral Ecology

Thanks to Bob Downs's comprehensive list of mineral species and Jolyon Ralph's massive compilation of mineral localities, DCO's dream of cataloging all of Earth's crystalline forms of carbon lies within our grasp. What does all that information tell us? That's the question we asked in 2014 when we began to look for hidden patterns in the mountains of data.

The first thing that struck us was the skewed distribution of Earth's minerals—a pattern observed commonly in biological ecosystems. A few scores of minerals occur at thousands of localities, while most minerals are extremely rare. A half-dozen species of the common feldspar mineral group account for an estimated 60 percent of the crust's volume.[37] A few dozen other common minerals make up almost all of the rest. By contrast, more than 1,200 mineral species are known from only one unique locality worldwide. Another 600-plus species have been found at exactly two sites, while almost 400 species are described from just three localities. A comprehensive survey of locality data in Mindat.org reveals that more than half of all documented minerals come from five or fewer localities. The striking conclusion: most mineral species are very rare.

So we began to wonder, Is that kind of skewed "frequency distribution," with a small number of very common types and far more numerous rarities, seen elsewhere in nature?[38] Might the literature of sociology, economics, geography, or other fields show an analogous frequency distribution? Is there an established mathematical approach to describing such a rarity-dominated relationship—one that we could exploit to understand minerals and their distribution?

The answer came, as answers often do, in a seemingly unrelated context—a walk in the woods. In June of 2014, the Carnegie Institution's newly appointed president, Matthew Scott, invited me to visit his Palo Alto home to discuss science, life, and the future of the institution. Matt is a kindred spirit, with a passion for broad concepts and multidisciplinary thinking. He has made his own revolutionary contributions to cellular and developmental biology, and he led the ambitious Bio-X lab at Stanford University, which gathers researchers engaged in forefront cross-disciplinary projects that link biology with medicine, engineering, physics, and chemistry. A billion dollars' worth of funding flowed through the state-of-the-art facility under Matt's leadership. Now he was anticipating a new adventure, guiding the renowned Carnegie Institution with its intertwined portfolio of exploration in Earth, space, and life sciences.

Rather than sit and talk, we went for hikes along the picturesque, rocky, northern California coast and in a nearby redwood forest, with its stand of ancient, massive trees. Walking past the imposing conifers, I was struck by the nonuniform distribution of plant and animal life. Most of the ecosystem's biomass was stored in the giant redwood trees, while most of the rest resided in a few other big, dominant species of trees and shrubs. But the vast majority of the biodiversity was to be found in much smaller species: mosses, ferns, insects, singing birds, and colorful California banana slugs, not to mention countless unseen microscopic life-forms. As I walked, I wondered, Could the distribution of biomass in an ecosystem mimic the distribution of minerals on Earth?

The breakthrough came from an unexpected source a few days later, stumbled upon while I was searching for articles on frequency distributions.[39] The answer was words. It turns out that the characteristic distribution of words in a book is remarkably similar to that of minerals on Earth. Consider this book. Like everyone else, I use words like "a," "and," and "the" a lot—probably hundreds of times each. Other frequently used words are more appropriate to this particular story; "mineral," "diamond," and "carbon" come to mind.

You've probably seen "word clouds" or "Wordles" that highlight the most common keywords in a text. What you don't see in a word cloud are all the unusual words—the ones used just once or twice. But many more, different words fall into this category. In this book "Wordle" appears only once (oops, I guess now it's twice). The same is true for "Chaucer," "tilapia," and "widgiemoolthalite." In fact, analysis of those rarer words can point an unambiguous finger to the subject matter, the genre, and even the authorship of a document. Have you found an old manuscript and want to know who wrote it? Those rare, idiosyncratic words and phrases might reveal a previously unknown work of Dickens, Chaucer, or Shakespeare.

Such a pattern, with a few common elements and numerous rare ones, is called a "Large Number of Rare Events" (or LNRE, for short) distribution. You might think that the study of LNRE distributions would be the purview of some obscure backwater of applied mathematics, of interest to only a few historians and literary scholars. On the contrary, "lexical statistics" is now a hot topic, thanks to the global war on terrorism. The National Security Agency wants to know who is writing what to whom. LNRE analysis, even of an email, short document, or transcript of a telephone conversation, can provide compelling clues. Consequently, money has flowed to LNRE research. Thick textbooks dense with mathematical formulas have appeared in recent years, while sophisticated statistical packages for LNRE analysis are available free online.

Delving into the nitty-gritty of such complex mathematics is not for the faint of heart, and few mineralogists have the know-how to decipher the arcane LNRE equations, much less customize them to a new discipline. In 2015, I was lucky to team up with Grethe Hystad, at the time an applied mathematics instructor at the University of Arizona and a onetime hockey teammate of Bob Downs. Finding Hystad was a scientist's dream.[40] She is mathematically gifted, eager to learn, creative, and as hard a worker as you'll ever find.

Norwegian by birth, Grethe has a pedigree that stretches back to the time of the Vikings. She spent much of her childhood on the farm that has been in her family for sixteen generations, and she boasts that the cache of Iron Age jewelry found on their land is a national treasure. Grethe is an avid athlete, having played for a Division 1 Norwegian soccer team, and having carried the Olympic torch for the 1994 Lillehammer Winter Games before coming to the United States for graduate studies. She completed her doctorate at the University of Arizona and stayed on as a lecturer in the math department before taking a professorship at Purdue University Northwest.

Grethe loved the concept of applying a well-established mathematical formalism to a new natural system: the distribution of minerals on Earth. She immersed herself in the literature of lexical statistics, extracted and modified the relevant procedures, and quickly demonstrated that the natural distribution of minerals on Earth conforms beautifully to two well-known types of LNRE distributions: the "finite Zipf-Mandelbrot" (fZM) and the "generalized inverse Gauss-Poisson" (GIGP) distribution functions.[41]

A rush of discoveries followed in a field we dubbed "mineral ecology," in homage to ecological studies of the distribution of species.[42] We found that LNRE distributions also apply to various subsets of minerals—notably, those containing specific chemical elements, such as boron, cobalt, copper, and chromium. Detailed studies of carbon took this idea a step further, revealing LNRE distributions for smaller subsets of minerals containing carbon in combination with oxygen, hydrogen, and calcium.

LNRE models of mineral distribution are intriguing in that they suggest an empirical law that accurately models what we had deduced from big mineral databases—that most minerals are rare. But there's much more to be gained from this approach. Mathematical models are invaluable not just because they systematize what we already know. These relationships also often take us beyond the mere description of nature, beyond what we know, to make predictions about what we don't know. Grethe Hystad's revelation was that LNRE models not only quantify the distribution of known minerals, but also reveal the distribution of minerals not yet discovered and described. With the LNRE model, we can predict Earth's "missing minerals."[43]

Here's how it works. Imagine landing as a space voyager on an unexplored Earthlike planet, where you are tasked with making as complete a mineral list as possible. The first rock you pick up will have several species new to your list. Pick up a different rock, and then another and another. As you seek out novelty, you will quickly expand the list. But a few weeks later, after you've cataloged thousands of samples and hundreds of different kinds of minerals, the discovery of new species will be less frequent, eventually slowing to a trickle of odd, rarer finds.

As you graph the growing number of mineral specimens examined on the horizontal axis versus the number of different species described on the vertical axis, you will see a characteristic "accumulation curve" that starts rising steeply on the left and gradually levels off to the right. From that curve you can extrapolate far to the right to estimate the total number of species, many of which have yet to be discovered and described. It will undoubtedly take many years of searching to approach, much less attain, that predicted number, but you can be confident that many more minerals lie in wait for the sharp-eyed mineralogist.

The LNRE formalisms provide a slick means to calculate accumulation curves, which Grethe Hystad generated from LNRE statistics with a few mathematical tricks thrown in. Our first collaborative effort, published back in 2015, when there were only about 4,900 known mineral species, predicted that at least another 1,500 minerals remained to be found. Subsequent research by a growing team of graduate students, postdocs, and senior scientists has zeroed in on details of what's missing: more than 100 minerals containing the useful element boron are just waiting to be found and described. We predict that 30 chromium minerals are missing, along with another 15 minerals of the scarce element cobalt. Studies on many other chemical elements followed, all based on the analysis of statistical trends in mineral data.

We saved our most exhaustive study for the more than 400 minerals of carbon, with almost 83,000 Mindat.org data available on these varied carbon minerals and their localities.[44] The LNRE distribution fits beautifully, with slightly more than 100 carbon minerals known from only one locality, another 40 known from exactly two localities, and so forth. The resulting accumulation curve suggests the tantalizing prospect that almost 150 carbon-bearing minerals exist at or near Earth's surface but have yet to be discovered and described. Pushing these methods further, we found that of those 150 missing minerals, almost 90 percent are likely to incorporate the most common mineral-forming element, oxygen, and almost as many incorporate hydrogen. We predicted that dozens of not yet discovered carbon minerals also embrace the elements calcium or sodium as essential building blocks.

With this information in hand, it was relatively easy for us to take the next step and predict both the identities of specific unknown minerals that might be found and where to look for them. Some of these

potential species are already well known as synthetic compounds—
carbonates of sodium and potassium, for example. Those chemicals are
usually white or gray and poorly crystallized, not to mention soluble
in water, so they disappear after every rain. Little wonder, then, that
those mineral species are likely to have been overlooked by amateur
and professional mineralogists. Our suggestion: go hunt for new species along the mineral-encrusted shores of sodium-rich Lake Natron
in the East African Rift Valley of Tanzania. It won't be an easy task,
because the lake's margins are already thick with more common white,
crusty minerals, but it's a lot easier to find something new if you know
what you're looking for.

We can guess the identities of other missing minerals by considering chemical cousins of known carbon minerals. Our list of 190 possibilities, such as iron, copper, and magnesium analogs of well-known
carbonate mineral species, just scratched the surface of plausible missing carbon minerals. With mineral ecology, we had transcended our
core Deep Carbon Observatory mission of finding all the forms of carbon on Earth; for the first time in mineralogical history, we predicted
a host of mineral species just waiting to be discovered.

And so we staked our claim. We proposed that Earth holds almost
150 missing carbon minerals, with explicit predictions about where to
go and what to look for. It was time to put the predictions to the test.

The Carbon Mineral Challenge

For centuries, mineralogy has been an observational science, with each
new mineral find a serendipitous thing. The rare sodium mica wonesite
was discovered by chance during routine analyses of common biotite.
The fibrous mineral jimthompsonite was long mistaken for ubiquitous
amphibole species. And, as the saying goes, gold is where you find
it. Sure, some rules apply, but only a handful of the more than 5,000
mineral species have been predicted prior to their discovery in nature.

With mineral ecology, that tradition is changing. We can predict
what's missing. We know what some of those rarities must be and

where to find them. We've seen that Tanzania's Lake Natron is the place to go to find new sodium and potassium carbonates. Similarly, a number of strontium carbonate minerals have already been found in Quebec's famed Poudrette Quarry, while other similar strontium carbonates are known only as synthetic chemicals.

To find new strontium carbonate minerals, you don't have to visit the Canadian quarry (though such an expedition would be a treat for any card-carrying mineralogist). Just go to museum drawers filled with specimens from Poudrette and scrutinize them for tiny crystal grains of previously unrecognized species. And there must be new carbon minerals in coal and oil shale, as well. Research has already yielded a dozen uncommon crystals formed from small carbon-bearing "organic" molecules concentrated in crystal-rich pockets of coal or layers of oil-rich shale. Surely, more organic minerals await discovery. To find them, you can dissect, examine, and analyze coal and oil shale from locations that have already yielded unusual minerals.

To promote this new mineralogical enterprise, the Deep Carbon Observatory initiated the Carbon Mineral Challenge in 2016.[45] This international quest for the missing minerals of carbon seemed like a fun idea, but we needed a charismatic leader—someone to build a global sense of excitement and mission. We needed someone who can talk to mineral curators and collectors alike. Meet Dan Hummer.[46]

He's hard to miss. It's not just his 6-foot 5-inch height and solid frame to match. Dan radiates enthusiasm with a spontaneous smile and an unpretentious kindness and generosity that set him apart. And there's a contagious "Aw, shucks" kind of wonder and enthusiasm, perhaps a vestige of his Iowa roots coupled with an ingrained curiosity. When Dan Hummer says carbon minerals are just waiting to be discovered, everyone nods and gets to work.

Dan, my former postdoc and a recently minted assistant professor at Southern Illinois University, understands what's at stake for the Deep Carbon Observatory. Our success depends on understanding Earth's complex carbon cycle, and we can't understand that cycle without knowing the myriad beautiful and varied forms that carbon assumes. A shortfall of almost 150 carbon-bearing minerals is a huge gap in our understanding of the natural forms of Element 6—one that Dan is determined to fill.

Mineralogists, both amateur and professional, from around the world have joined the hunt, and the results are pouring in. Within the first year of the challenge, the IMA approved nine new carbon-bearing minerals as valid species. Abellaite was the first to be found—a sodium lead carbonate with tiny sprays of pale-green needles from Catalonia, Spain. We were delighted to note that abellaite, approved in 2017, was in our 2016 published list of predicted carbon minerals. The second find, tinnunculite, forms when the excrement of falcons (species *Falco tinnunculus*—hence the name) interacts with hot gases from a burning Russian coal mine.[47] (OK, I admit we didn't predict that one!) Blue marklite from Germany, green middlebackite from Australia, and pale-yellow leószilárdite from Utah followed. Lovely canary-yellow ewingite, the sixth discovery, is a new uranium carbonate from the Jáchymov ore district of the Czech Republic, a locality already known for its diversity of rare carbon minerals. And the eighth new carbon mineral, parisite, a carbonate with the rare element lanthanum, was another predicted phase.

The Carbon Mineral Challenge runs through 2019. We don't expect to find all of the remaining 145 predicted new species, but, as Dan Hummer promised, we'll sure have fun trying.

—

It's only natural that the vast majority of nature's carbon-bearing crystals have been discovered in the accessible near-surface realm of Earth's crust. But we know that Earth holds deeper mineralogical secrets—crystals hidden from view, forged at the extreme temperatures and pressures of our planet's mantle and core. Understanding those mysterious phases requires a sophisticated arsenal of research tools operated by a special breed of scientists. Enter the mineral physicists.

DEVELOPMENT—Deep Earth Carbon

HUNDREDS OF MILES beneath Earth's solid surface lies a hidden, inaccessible realm of mystery. Extreme pressure and temperature, conditions inimical to life, are the twin forces that shape the deep interiors of planets. Atoms crush together and collide, adopting exotic, denser crystalline forms. Our view of the Cosmos is warped by our existence at the all but-impenetrable boundary between Earth and Air. We are constrained to walk on Earth's solid surface—a rocky barrier that thwarts our efforts to explore anywhere but the thinnest veneer of our majestic planetary home.

What astounding discoveries await a hundred, a thousand miles beneath our feet?

Deep Carbon Mineralogy

Our inventories of carbon-bearing mineral species, however comprehensive, only scratch the surface; that is, literally the surface—the accessible upper mile or two of Earth's crust. Almost all known mineral species grew and were recovered in that thin, rocky shell. And many minerals—those collected from weathered mine dumps or formed from smoldering falcon poop—are as shallow as they come.

At the Deep Carbon Observatory, we are greedy to know more.

We want to understand the hidden, inaccessible, deep realm of Earth's mantle and core, where immense pressures and temperatures crush and sear carbon and its companion elements, coaxing them into novel, dense forms that are only gradually coming to light. We must learn those tantalizing secrets of the deep, for almost all of Earth's carbon is locked away inside the planet. For us, Earth is like an immense spherical jigsaw puzzle with only a few edge pieces securely in place. We are eager to fill in the missing pieces of the carbon mineral puzzle, but there's a big hurdle: the deeper we go, the more challenging the problem becomes.

—

Of the more than 400 known carbon minerals, no more than a handful of species represent high-pressure phases.[48] Diamond, forged at the extreme temperatures and pressures of Earth's deep interior, is the most obvious example of a carbon-bearing mantle mineral. Dense moissanite, a "carbide" compound with carbon atoms bonded directly to silicon in a diamond-like crystal structure (a structure notably lacking oxygen), is another likely candidate. Because silicon carbide crystals possess physical properties remarkably similar to those of diamond, faceted and polished synthetic moissanite gemstones find a ready market as a relatively inexpensive diamond substitute. Evidence from rare inclusions in diamond points to a few other possible mantle carbide minerals in which carbon atoms bond to the metals iron, chromium, or nickel. But that's about it for natural specimens from the deep. What else might be down there?

The standard protocol for identifying plausible mantle minerals is to subject common crustal minerals to the unforgiving conditions of a hundred miles or more beneath Earth's surface. Common calcite, the ubiquitous carbonate of calcium, was one obvious mineral species to try. I vividly recall reading the pioneering work of William (Bill) Bassett and his graduate student Leo Merrill, who described the first of a sequence of dense, high-pressure forms of calcite.[49] I was myself a fledgling graduate student looking for a good thesis project. Bill had a seductive answer: high-pressure crystallography.

To a scientist like Bill Bassett, "deep carbon" translates to "high-pressure carbon." The deeper you go into Earth's interior, the higher the pressure. Earth's mantle subjects minerals to pressures of hundreds of thousands of atmospheres, while pressures at the core exceed 1 million atmospheres. It was challenging enough for Scotsman James Hall to mimic conditions a mile down in his daring gun barrel experiments. To match the environment of Earth's mantle is as intractable an experimental challenge as you're likely to find.

An added difficulty for an experimentalist studying crystals is to generate Earth's extreme pressures without crushing a crystalline sample to powder. It's a game of trade-offs. You want the highest possible pressure, which requires subjecting tiny areas to big forces. But tiny areas mean tiny crystal samples that are easily crushed. How do you measure small crystals at pressure without destroying what you want to probe? The problem is compounded because your pressurized sample has to be enclosed in a strong, protective chamber. How is it possible to make any useful measurements through such a solid barrier?

A brilliant solution to this experimental challenge emerged from the US National Bureau of Standards (NBS) in the 1950s, when NBS scientists enjoyed an unexpected opportunity to investigate diamonds. They had been given a large cache of cut diamonds seized from smugglers and were told they could perform just about any experiments they wanted. One batch of valuable stones, hundreds of carats of brilliant-cut treasures, was burned away to nothing in a futile search for impurities (the answer: there aren't many impurities in gem diamonds). Other diamonds, including one magnificent 8-carat gem worth a small fortune, were scratched, drilled, or shattered.

It was during those abuses that NBS scientist Alvin Van Valkenburg realized the unmatched potential of diamonds to serve dual roles in the high-pressure game—both as a strong pressure vessel to confine and squeeze a sample, and as a clear window into the world of that compressed sample. Van Valkenburg matched pairs of brilliant-cut gem diamonds, each with their pointy ends flattened off, to focus pressure in the "diamond-anvil cell," or DAC.[50] His simple viselike design squeezed the diamonds together to generate huge pressures while protecting the sample crystal.

You assemble the DAC's sample chamber layer by layer. The bottom layer is a flat, steel plate with a small cylindrical hole drilled through it. Take the first diamond anvil and center it over the hole, anvil end up. The next layer is a "gasket," cut from a thin metal sheet no more than two-hundredths of an inch thick. A small hole in the gasket, precisely centered over the lower diamond, serves as the cylindrical walls of a sample chamber. Load that chamber with three ingredients: first your crystal sample (usually anchored in place by the smallest dab of Vaseline), then tiny grains of pressure-sensitive ruby or some other material to serve as an internal pressure standard, and finally, to top off the sample chamber, water or some other pressure-transmitting fluid. The chamber seals when the second diamond is positioned on the gasket, its centered anvil face pointing down and backed by a second steel plate. Once the sample chamber is assembled, you pressurize it by squeezing it in any of a variety of viselike devices. If you've been careful and all the cylindrical holes line up, you can look straight through the diamonds into the amazing, unexpected world of high pressure.

The NBS team made high-pressure history with their new "toy," as they called it. They watched, amazed, as pure water transformed to new high-pressure forms of ice, and alcohol crystallized to bladed needles—what Van Valkenburg dubbed "gincicles." They used fancy spectrometers to measure dramatic changes in light's interactions with matter. And they shot beams of X-rays at their samples in an effort to glimpse the ways that atoms rearrange themselves, adopting denser configurations when squeezed.

I was captivated by the stunning reports of Van Valkenburg and his NBS colleagues. When I read their breakthrough papers in the early 1970s, and caught that tantalizing glimpse into the previously hidden deep realm, I knew that's what I wanted to do for a living.

X-Raying Crystals at Pressure

When DCO scientists talk about discovering all the varied "forms" of carbon, we have a very particular mental image in mind. We imagine atoms. All the materials around you—all the solids, liquids, and gases—are made of atoms. Crystals, with their elegantly repeating symmetrical patterns of atoms, hold special appeal. Each mineral species has its own distinctive atomic topology, its own "crystal structure."

Pressure adds a wrinkle to the crystal-structure story. Subject a mineral to higher and higher pressures and its atoms must snuggle closer together in denser and denser arrangements. If we are to understand Earth's deep forms of carbon, then we must discover those dense, high-pressure crystal structures.

X-ray diffraction provides the elegant means to measure atomic structures of crystals. X-rays are an energetic form of light waves, similar in character to visible light and radio waves, but with much shorter wavelengths, a few billionths of an inch—distances close to the regular spacing between layers of atoms in crystals. When an X ray beam bathes a crystal, the waves are scattered and reinforced in focused sprays of "diffracted" X-rays. The directions and intensities of those diffracted X-rays reveal the atomic structure.

The original NBS diamond-anvil cell was an amazing advance, but the prototype design was much too bulky to mount in a standard beam of X-rays. What's more, the steel support system of the NBS design blocked most incoming X-rays. Merrill and Bassett's elegant solution, published replete with mechanical drawings in the prestigious *Review of Scientific Instruments* in 1974, was to construct a miniature version of the diamond-anvil cell with X-ray-transparent beryllium metal replacing the steel backing plates.[51] A triangular frame with three screws provided the squeezing force of the Merrill-Bassett cell.

Their first experiments focused on calcite, which was known to undergo transitions to slightly denser atomic arrangements known as "calcite-II" and "calcite-III" at pressures close to 15,000 and 20,000 atmospheres, respectively—pressures found in Earth's uppermost mantle a few tens of miles beneath our feet. Merrill and Bassett weren't able to decipher all the details of these structures, but they did show small distortions in the atomic arrangements that pointed to denser forms with lower crystal symmetries.

Eager to try my hand at this new approach and apply it to my PhD thesis, I contacted Bill Bassett and asked for his advice. Some scientists would have balked. Armed with a powerful new technique and a host of choice problems to tackle, why encourage competition? But Bill went out of his way to help. He had his machine shop make a new diamond cell just for me, sold it to me at cost, and made the trip from Rochester, New York, to Cambridge, Massachusetts, to show me how to use it.

Bill Bassett also helped a lot of other scientists, and the field of high-pressure crystallography flourished. Thanks to Bill's pioneering efforts, calcite continues to be a source of fascination. At least six different forms of calcium carbonate are now known to occur at pressures up to about 80,000 atmospheres, all of which incorporate the typical tiny carbonate triangle with three oxygen atoms neatly surrounding a carbon atom. The carbonates of iron, magnesium, manganese, and other elements display similar diversity of forms at pressures equivalent to those of the upper mantle—pressures reached relatively easily in DAC studies of crystal structures. As a consequence, we now know that the minerals of Earth's deep interior are different from those in near-surface realms.

Higher Pressures

Probing crystal structures at the extreme conditions of Earth's transition zone and lower mantle, where pressures exceed 100,000 atmospheres, presents new challenges. One successful strategy has been to apply theories of atomic bonding to the problem. Advances in quantum mechanics have led to sophisticated mathematical models of crystal structures—computational techniques that accurately reproduce many of the structures of natural materials found in Earth's crust, as well as those of synthetic compounds, some of which were predicted by theory before they were synthesized in the lab.[52]

It takes some mathematical tricks and powerful computers, but these computational methods can also be extended to high pressure and temperature. Computer models have once again proved remarkably successful, replicating known high-pressure transitions to dense mantle minerals (even though they don't always predict the exact pressures at which new mantle minerals emerge). Unlike experiments, where every pressure increment adds layers of experimental complexity, it's a fairly simple matter to dial in a million atmospheres or more of pressure in a quantum calculation and see what happens.

The universal result, not surprisingly, is that deeper minerals adopt denser structures. For carbonate minerals like calcite and dolomite, that initial increase in density arises from progressively tighter packing of the familiar carbonate (CO_3) triangles with other atoms, but repackaging atomic triangles and balls can get you only so far. Above about a half-million atmospheres, we need a different strategy, so carbonate minerals take a page from diamond. The graphite-to-diamond transition features carbon switching from three neighboring carbon atoms in a plane, to four neighboring atoms in a pyramid. In like manner, computations suggest that carbon in carbonates will shift from having three neighboring oxygen atoms in a plane to having four oxygen atoms in tiny pyramid-shaped CO_4 groups called "tetrahedra."

Mineralogists quickly recognized the possible similarities of CO_4-bearing high-pressure carbonates to the many common silicate minerals found abundantly in Earth's crust—minerals that incorporate silicon surrounded by a tetrahedron of four oxygen atoms. Dozens of familiar silicate structure types—mica, feldspar, pyroxene, garnet, and more—dominate Earth's crustal mineralogy. Could similar structure types occur in mantle carbonates? Sure enough, theorists predicted that magnesium carbonate at lower mantle conditions should adopt the elegant pyroxene structure, with long chains of CO_4 tetrahedra linked corner to corner.

In spite of these intriguing predictions, most geophysicists want experimental proof—verification that requires X-ray crystallography at seemingly impossible extremes. The technical advances required for such research are daunting. In the 1970s and 1980s, our state-of-the-art methods employed "big" crystals, a hundredth of an inch across, with a simple Merrill-Bassett DAC and the kind of conventional X-ray source that's available in any crystallographic lab. On a good day, we could approach 100,000 times atmospheric pressure without destroying our sample or cracking the expensive diamond anvils. Those high-pressure X-ray experiments eventually became routine and were replicated in dozens of labs around the world.

It's a different story at hundreds of thousands of atmospheres. Crystals must be less than a thousandth of the volume of the crystals I examined; any larger and they would be crushed to powder. A much more sophisticated DAC design must be employed, lest misaligned diamonds crack and shatter at the much higher pressures. Conventional X-ray beams are also inadequate; they're much too weak to generate measurable patterns from a crystal smaller than a speck of dust. So, scientists have to travel to giant government-run "synchrotrons," particle accelerators in which X-ray beams a million times more intense than conventional sources are in high demand twenty-four hours a day, seven days a week. Few scientists have mastered these exacting constraints, which provide the only experimental path to understanding Earth's deepest carbon-bearing crystals. Among them, mineralogist and crystallographer Marco Merlini of the University of Milan in Italy stands out for his discoveries in deep carbon science.[53]

Marco Merlini is a modest scientist who seems much more enthusiastic about the thrill of discovery than about any personal recognition. He greets you with a ready smile and eager eyes, itching to show you his lab and his latest results. And those results are spectacular. In a 2012 paper published in the US *Proceedings of the National Academy of Sciences*, Merlini and colleagues reported on high-pressure structures of dolomite, a common crustal carbonate with equal proportions of calcium and magnesium and a structure similar to that of calcite.[54]

If carbonate minerals exist in Earth's mantle, then dolomite is a good candidate. Working at the European Synchrotron Radiation Facility in Grenoble, France, Merlini's team compressed a tiny dolomite crystal to unprecedented extremes. Above 170,000 atmospheres, the researchers found a structure they dubbed "dolomite-II" that is similar to Merrill and Bassett's calcite-II. Squeeze to 350,000 atmospheres and a completely new structure appears—one with four oxygen atoms around some of the carbon atoms, but in a novel, flattened pyramid; they called it a "3 + 1" configuration. Merlini and his team continued to explore their dolomite crystal to 600,000 atmospheres, but they did not see a sign of transformation to the postulated carbonate with carbon surrounded by a tetrahedron of oxygen atoms.

The breakthrough came in 2015, when Merlini's group published the description of a remarkable new high-pressure form of a carbonate containing equal parts of magnesium and iron.[55] The measurements seemed all but impossible, requiring pressures approaching a million atmospheres, equivalent to conditions in Earth's deepest mantle more than a thousand miles beneath our feet. The work of these researchers confirmed the predicted transformation of flat CO_3 carbonate groups into pyramids of CO_4. Yet, rather than adopting the expected pyroxene structure, with continuous chains of corner-sharing tetrahedra, they found a completely new and unexpected atomic arrangement. Their ultra-high-pressure carbonate had broken chains with segments of four tetrahedra separated by short, iron-filled gaps—a wonderfully quirky, dense structure unlike anything seen before.

Merlini's discoveries have profound implications. Decades ago, conventional wisdom held that high-pressure minerals would tend to have simple structures—a consequence of the necessary dense, regular packing of atoms at depth. The expanding body of research by Merlini and other ultra-high-pressure pioneers tells a different story. High-pressure structures can be complex, novel, and often unexpected. And that's good news for those of us who thrill in nature's surprising complexity.

Deep Diamonds[56]

Of all the varied high-pressure forms of carbon-bearing minerals, including crystalline forms both known and yet to be discovered, diamond will always hold pride of place. Diamond occupies the ideal niche between scarce and rare: it is sufficiently abundant that almost everyone can own one, but rare enough to command millions of dollars for newsworthy large stones. Hundreds of millions of gems large enough for a ring or necklace have been mined, but hundreds of millions of consumers want to possess one or more. The lure of diamonds extends to their scientific value; the more we study these almost pure fragments of carbon from Earth's depths, the more we learn about the history and dynamics of our planet. Little surprise, then, that no other mineral species has so mesmerized the scientists of the Deep Carbon Observatory.

The universal process of carbon atoms condensing from hot gas in the envelopes of energetic stars produced the first diamonds—the first crystals (albeit at a microscopic scale) in the history of the Universe. But that energetic process, conducted in the near vacuum of space, is not the way our most prized diamonds form. For gemstones, we must shift our gaze from the outskirts of stars to the deep interiors of planets like Earth.

Earth's crust manufactures large amounts of graphite. When carbon atoms concentrate anywhere near the planet's surface, it is graphite that appears, not diamond. To craft a big crystal of dense, hard diamond, Earth needs a great deal of pressure—at least tens of thousands of times atmospheric pressure—to pack carbon atoms closer together. It doesn't hurt to apply the heat of a blowtorch as well, to coax wiggling carbon atoms into their new, more stable pyramidal configuration. And so we must shift focus to the deep interior, a hundred miles or more down into Earth's inaccessible mantle. There, when chemical conditions are just so, when the pressure and temperature are high enough, and when lots of carbon atoms concentrate and nucleate, is where prized gemstones can grow.

Humans have learned to mimic this process by constructing giant hydraulic presses with tough carbide anvils and powerful electric heaters to replicate conditions hundreds of miles beneath our feet. Millions of carats of synthetic stones are manufactured this way each year—some for abrasives, some for electronic components, some for optical windows, and some as synthetic gemstones. You can even commission a wearable "memorial diamond," forged at pressure from carbon atoms collected during a loved one's cremation. People aren't forever, but a memorial diamond will outlast even the most deeply entrenched memory.

The Largest Diamonds Are Different[57]

Now that we know diamonds can reveal long-hidden secrets of our planet's complex deep interior and dynamic past, a growing scientific community is finding new reasons to value diamonds above all other gemstones. The diamond seekers of this new generation do not crave the flawless stones of high-end engagement rings and tennis bracelets. On the contrary, above all else they prize imperfections in the form of tiny mineral inclusions—unsightly black, red, green, and brown mineral specks, and microscopic pockets of deep fluid and gas. These blemishes, typically cut away and cast aside in the faceting of precious gemstones, often represent pristine fragments of Earth's deep interior—bits and pieces that originated long ago, far below our planet's sunlit surface, where they were trapped and hermetically sealed as the growing diamonds engulfed them.

The stories they tell! Diamond inclusions have the potential to divulge how deep, how long ago, and in what surroundings the diamonds grew.[58] Consider secrets that are now being revealed by the world's largest stones. In the rich lore of diamonds, giant gems stand out: the 603-carat Lesotho Promise, unearthed in 2006 and touted as the greatest find of the new century; the legendary 793-carat Koh-i-Noor diamond found centuries ago in India, now in the crown of the British Queen Mother; the 813-carat Constellation, sold at auction in

2016 for a record $63 million; and the most outsized treasure of them all, the 3,106-carat Cullinan, which was discovered in 1905 at South Africa's Premier Mine as the surviving fragment of what must have been a much larger stone. It turns out that all of these giants share a common, unexpected origin.

For centuries, it was assumed that such magnificent gems are just large versions of more common, smaller stones. Not so. Hints at a different genesis come from optical studies. Most diamonds, though stunningly transparent to visible light, absorb wavelengths of infrared and ultraviolet light as a consequence of impurities at the atomic scale. Nitrogen atoms are the most common offenders. In "Type I" diamonds, nitrogen typically replaces about one in every 1,000 carbon atoms. When those nitrogen atoms congregate into little clusters, they may impart a yellow or brown color to the gems. Once thought unsightly, some of these impure crystals are now marketed under such cleverly seductive names as "cognac diamonds," "champagne diamonds," and "chocolate diamonds." Sorry, but they're still just brown diamonds.

The remaining diamonds, less than 2 percent of all mined gems, are "Type II." Distinguished by their unparalleled transparency to both visible and ultraviolet light, Type II diamonds have no discernible nitrogen impurities, and they tend to be larger and more optically perfect—characteristics that have led some scientists to posit a slower, deeper crystallization environment. Nevertheless, the exact origins of Type II diamonds remained a mystery.

In a headline-grabbing 2016 discovery, an international team of deep carbon scientists led by Evan Smith of the nonprofit Gemological Institute of America, or GIA,[59] showed that Type II diamonds, including many of Earth's biggest gemstones, host a distinct and curious suite of inclusions: silvery specks of iron-nickel metal quite different from the usual oxide and silicate mineral inclusions of their smaller cousins.

This research is a triumph on sociological as well as scientific grounds. Mine owners, gem cutters, and collectors jealously guard their hoards; the bigger the diamond, the more difficult it is to gain access for scientific study. To win the opportunity for even a cursory examination of inclusions in one or two big diamonds would be an unexpected treat for most scientists. Those who tried earlier, who caught brief glimpses of the silvery inclusions in big stones, mistakenly assumed them to be the common mineral graphite—a result that was not particularly newsworthy.

But Smith and his colleagues at the GIA, teaming up with other diamond experts from the United States, Europe, and Africa, laid the groundwork for studies at an altogether grander scale. The GIA in New York is tasked with certifying diamonds of all kinds: weighing them, grading them, teasing out their countries of origins, and constantly devising new tests to weed out the next generation of crafty synthetic fakes or illicit "conflict diamonds." GIA certification is the universal standard of excellence for diamonds. From numerous contacts at mines and museums, Smith's team was able to assemble and probe in detail an astonishing collection of gems and cutting fragments from fifty-three big Type II diamonds. They even recut and polished five of the fragments to expose the silvery inclusions to the meticulous probing of advanced analytical instruments.

The first surprise came from composition studies. The metal-rich inclusions contain no oxygen, the mantle's most abundant chemical element, but they are rich in carbon and sulfur—telltale impurities revealing that the metal must have been in a molten state when the diamonds formed. Remarkably, metal inclusions point to deep regions of our planet similar in composition to Earth's inaccessible core, with its ocean of dense liquid iron and nickel surrounding a 1,520-mile-diameter inner sphere of even denser crystalline iron and nickel alloy.

This is what we can infer: Big diamonds grow hundreds of miles beneath the surface in isolated mantle pockets of metal-rich liquid. Diamonds grow easily in such environments because iron metal has the unusual ability to soak up lots of carbon atoms. At sufficient pressure and temperature, diamonds nucleate and grow, with mobile carbon atoms passing easily through the metal melt, adding layer upon layer to potentially giant crystals. It's not a complete surprise to scientists that some diamonds form in this metal-mediated way; metal solvents have been employed to grow large crystals in the synthetic diamond business since the early 1950s. But no one realized that nature had learned the same trick billions of years earlier.

The implications of this finding, that big diamonds have their own special provenance, go far beyond the quest for fancy gems. This distinctive population of Type II diamonds reveals a previously undocumented heterogeneity in the mantle. One might think that the mantle's high temperatures, coupled with billions of years of mixing by convection, would have blended the mantle into a smoothie-like uniformity. Now, thanks to big diamonds and their telltale inclusions, we have clear evidence that the mantle is more like a fruitcake, with some relatively uniform regions, but with swirls of novelty and lots of fruits and nuts (read metal and diamonds) thrown in.

What's more, these local variations in the mantle's rocks and minerals point to deep regions with wildly different chemical environments. We have long assumed that the mantle is made almost exclusively of silicates, oxides, and other oxygen-rich minerals. That's what we typically see in the volcanic rocks called kimberlites that transport their trove of diamond gemstones to the surface and host the world's richest diamond mines. But metal inclusions point to other mantle zones that are devoid of oxygen—regions where different chemical processes (like growing really big diamonds) can occur.

As in so many facets of Earth's evolution, the closer we look and the more data we collect, the more complicated and fascinating the story becomes.

Diamonds Record Secrets of Earth's History

Metal inclusions, observed in only a small fraction of diamonds, appear to be the exception—rarities matched by the paucity of the large Type II diamonds in which they are found. Far more common, and the bane of jewelers seeking the most flawless cut and polished gems, are a suite of more common mantle minerals in Type I diamonds. For as long as humans have treasured perfect jewels, mineral inclusions have been a source of disappointment. Scientists adopt a different viewpoint, for mineral inclusions are themselves a treasure trove of data on Earth's deep interior.

Some of these inclusions reveal the ages of diamonds, with a few ancient stones clocking in at more than 3 billion years. The keys to determining a diamond's birth date are the rare, microscopic bits of sulfide minerals—shiny crystals smaller than a hair's breadth containing combinations of metal and sulfur atoms. These sulfide inclusions always incorporate a tiny amount of the rare element rhenium, which is a remarkably useful element if you want to know the age of a mineral.

Natural rhenium atoms come in two flavors. The stable isotope rhenium-185 constitutes about 37 percent of Earth's reserves; the remaining 63 percent consists of radioactive rhenium-187—an unstable isotope that can spontaneously transform to stable osmium-187, with about half of the rhenium atoms transforming every 41.6 billion years. Over time, the ratio of radioactive rhenium-187 to osmium-187 decreases as predictably as a ticking clock. It takes exacting sample preparation and ultrasophisticated analytical gear, but a skilled and patient scientist can tease an age from diamond by measuring the ratios of rhenium and osmium isotopes in a microscopic sulfide inclusion.

That extreme dating exercise pays tremendous dividends when combined with studies of other entombed mineral grains—typically, the more abundant oxide and silicate minerals that form most of Earth's mantle. Those characteristic assemblages sometimes reveal the extreme depths of diamond formation; in a few cases, unusually dense inclusions of oxide and silicate minerals point to an origin more than 600 miles beneath the surface, well into the mysterious and inaccessible realm of Earth's lower mantle. How diamonds can make their way to the surface from such great depths, how they survive such a journey, finding safe passage through hundreds of miles of seemingly solid rock without fragmenting or getting stuck or breaking down to another mineral, remains a largely unsolved mystery.

By whatever tortuous pathway they emerged from the deep, these diamonds have stories to tell about billions of years of global-scale changes.[60] Remarkable evidence, meticulously gathered in a pioneering 2011 study by diamond experts Steven Shirey of the Carnegie Institution for Science and Stephen Richardson of the University of Cape Town in South Africa points to a profound transition that occurred roughly 3 billion years ago.

Systematic studies of mineral inclusions in diamonds from around the world—gems from the most productive mines in Brazil and Russia, South Africa and Canada—reveal that younger diamonds often carry distinctive suites of grayish-green pyroxene and red garnet—a colorful mineralogical duo that points to a source from a rock called eclogite. Eclogite is significant because of its ancestry. This beautiful red-and-green rock arises through the high-pressure transformation of common basalt, a dark rock that crystallizes from magma pouring from thousands of miles of deep ocean volcanic ridges around the globe. As a consequence of this volcanism, basalt covers most of the ocean floor—almost 70 percent of Earth's surface. This continuous production of new basalt crust at ridges requires a matching loss of old basalt crust through deep and irreversible burial at "subduction zones." Far from the volcanic ridges, the crust's older, colder, denser basalt layers bend downward and sink deep into Earth's interior. Subduction thus completes the essential recycling of Earth's surface via plate tectonics.

As basalt subducts, it experiences ever-greater pressures and temperatures. At depths 30 miles or more beneath the surface, basalt transforms into denser varieties of minerals, including the red garnet and gray-green pyroxene found in some diamonds. This characteristic combination of eclogite inclusions led scientists to conclude that modern-style plate tectonics, with its active volcanic ridges and dynamic subduction zones, has been going strong for the last 3 billion years.

Other diamonds, including all of those with ages older than 3 billion years, tend to hold very different suites of mantle minerals. You can find lots of inclusions of yellow or brown olivine (the semiprecious gemstone known as peridot), as well as purple garnet, black chromite, and emerald-green pyroxene. This distinctive assemblage of mineral inclusions points to a much deeper mantle source within a rock called peridotite, which we think dominates Earth's mantle. This dense collection of mineral fragments never saw the surface, nor was it affected by subduction. The profound implication is that early Earth did not experience plate tectonics—at least, not the modern style of colliding and fragmenting continents and subducting basaltic crust that we see today.

The lesson is clear. Diamonds and their inclusions are truly scientific treasures, providing us with compelling evidence for one of our planet's greatest innovations: the emergence of plate tectonics when Earth was about 1.5 billion years old. And, in an ironic twist not lost on diamond researchers, some of those once-scorned gemstones loaded with unsightly inclusions are now being sold to mineral collectors at premium prices. Highly publicized scientific discoveries have fascinated the public and, in the process, created a demand for inclusion-rich specimens that has priced some researchers out of the market.

Carbon in the Core

Catching glimpses of carbon mineralogy in the mantle is hard enough, but it's a cakewalk compared to probing below the core-mantle boundary almost 1,800 miles beneath the surface. There, pressures soar to above 1 million atmospheres with temperatures exceeding 5,400 degrees Fahrenheit. Understanding the extent and nature of carbon in the core remains the single greatest unsolved mystery in estimating Earth's total carbon content.

The mineralogy of the molten outer core is easy. There are no crystals, so there are no carbon minerals. It remains to be seen, however, how much carbon might be dissolved in this molten zone of iron-nickel metal. At least two lines of evidence suggest that there could be a lot—perhaps much more than in all of the rest of the planet combined.

The first deep carbon clues came from the pioneering work of

Harvard geophysicist Francis Birch, a quiet modest man.[61] Birch's scientific discoveries were perhaps overshadowed by his central role in the construction and deployment of the atomic bomb known as "Little Boy." As a lieutenant commander in the navy during the Second World War, he supervised the bomb's assembly on Tinian Island in the western Pacific Ocean, and he oversaw the loading of the weapon into the Boeing B-29 Superfortress known as the *Enola Gay*.

When I took his geophysics course in the fall of 1971, the sixty-eight-year-old Birch was a gentle, engaged teacher. He covered all manner of basic geophysics, from Earth's layered structure to its significant heat flow and variable magnetic field. If he hadn't been so famous in the field—if we hadn't already been schooled in "Birch's law" and the "Birch-Murnaghan equation of state" since undergraduate years—we would not have known how much of the course material was based on his own transformative discoveries.

In Birch's most influential work, published in 1952 and a mainstay of geophysical thinking to this day, he combined data from seismology (the study of sound waves traveling through Earth) with materials science.[62] Birch realized that the velocity of a seismic wave is directly related to the density of the rock through which it passes. Using his model, he described Earth's interior with a new degree of detail and sophistication. Beneath the thin crust is the three-layered mantle, with significant density discontinuities at depths of roughly 250 and 410 miles—boundaries that separate the upper mantle, transition zone, and lower mantle. Birch suggested that the properties of these layers conform to increasingly dense assemblages of silicate minerals rich in magnesium, silicon, and oxygen. Decades of subsequent work by hundreds of scientists have served to add detail and nuance, but Birch's big picture still holds true.

A much sharper core-mantle discontinuity, representing a dramatically greater density contrast, occurs at the base of the mantle, some 1,800 miles below the surface. Other scientists had long described the core as a dense, metal-rich zone, with a liquid outer core extending down to 3,200 miles and a smaller crystalline inner core with a radius of 760 miles. Birch employed new data on the density of liquid iron metal and alloys at high pressure and temperature to amplify this view, noting that seismic velocities in the core implied a density significantly lower than that of pure iron-nickel metal. There must be, he argued, at least one lighter element in that molten layer; the outer core's iron and nickel atoms are mixed with as much as 12 percent of something else. Could a vast store of carbon be that missing element?

Birch was quick to recognize the potential uncertainties of his bold deep Earth model. In a humorous footnote that has become as famous as his geophysical findings, Birch provided a caveat:[63]

Unwary readers should take warning that ordinary language undergoes modification to a high-pressure form when applied to the interior of the Earth. A few examples of equivalents follow:

High-Pressure Form	Ordinary Meaning
Certain	Dubious
Undoubtedly	Perhaps
Positive proof	Vague suggestion
Unanswerable argument	Trivial objection
Pure iron	Uncertain mixture of all the elements

In spite of his warning, Birch's prediction of a light element in the liquid outer core has withstood every test. But what could that element be? A devoted band of experimentalists and theorists has tackled that intriguing question—one that remains open to this day.

In seeking the answer, we must follow three simple ground rules. First, the element must be something significantly lighter than iron or nickel, so uranium, lead, or gold need not apply. Second, the element has to occur abundantly in the Cosmos; this requirement eliminates light lithium, beryllium, and boron, for example. Finally, the element must be able to dissolve in a metal melt at the outer core's extreme conditions of temperature and pressure. In fact, only a handful of candidates meet these three basic requirements—hydrogen, carbon, oxygen, silicon, and sulfur being the only viable contenders. Each has merits and flaws; each, its proponents and detractors. Of course, this is not an either/or proposition. A metal melt could easily dissolve more than one light impurity element, perhaps even all five at once. (This ecumenical solution is my preference, since nature seems to promote complexity.) In any event, there's strong evidence that carbon is in the mix.

Carbon isotopes provide convincing clues.[64] Carbon atoms come in two common varieties, two different stable isotopes. Every carbon atom has 6 protons in its nucleus; that's the definition of carbon. However, the number of neutrons, the other building block of atomic nuclei, can vary. Almost 99 percent of carbon atoms have 6 neutrons (that's "carbon-12"), while the remaining 1 percent is "carbon-13," with 7 neutrons. Nearby rocky worlds, including the red planet, Mars, and the big asteroid Vesta, share with each other a well-documented ratio of those isotopes—a ratio that seems to characterize most of the other bodies in our inner solar system. But Earth's carbon—at least the accessible carbon near the surface—seems to be too "heavy" by comparison, with a higher percentage of carbon-13 than its planetary neighbors have. That's a puzzle that begs for a solution.

The easiest explanation for this seeming anomaly is that Earth's isotopic composition is the same as that of other worlds, but the "missing" light carbon is hidden from our view, locked away in Earth's core. If the liquid outer core holds even a tiny fraction of carbon, then the core could easily sequester 100 times all the known carbon in Earth's crust. How much carbon does Earth hold? It is shocking how ignorant we are regarding such a profound question.

The Deepest Mysteries

No zip code of Earth is more remote, more impenetrable, than the solid inner core. Lying more than 3,200 miles beneath our feet, elements of the inner core are subjected to pressures above 3 million atmospheres and temperatures soaring to 9,000 degrees Fahrenheit. For decades, the conventional wisdom has been that crystalline iron metal with a dollop of nickel forms the inner core. As with the molten outer core, one or more light elements might also play minor roles, but iron is the leading actor.

But there's a problem—one rooted in the nature of sound waves. Seismic waves come in two different forms. The stronger, faster, primary (or "P") waves occur when atoms and molecules bump against each other in succession, like a wave of falling dominos. The motion of the atoms is in the same direction as the motion of the P wave. Iron and its nickel-bearing alloys successfully match the observed velocity of P waves in the inner core.

By contrast, secondary (or "S") waves occur when atoms move side to side, triggering a similar side-to-side motion in their neighbors. (Think of doing "the wave" in a football stadium, where people stand up and sit down but the wave goes around the stadium.) The motions of the atoms are perpendicular to the wave's motion. Surprisingly, S waves of the inner core travel only about half as fast as they should in crystalline iron.

What's going on? One simple explanation posits that the inner core is partially molten—a condition that invariably slows down S waves—but iron-nickel metal should not melt under presumed inner-core conditions. Jie "Jackie" Li, professor of geology at the University of Michigan, devised a clever alternative experimental solution to this discrepancy.[65]

Brilliant, engaged, quick to praise intriguing new ideas or point out flaws in a colleague's arguments, and equally quick to smile at a subtle joke or clever turn of phrase, Li is a master of the diamond-anvil cell. Like so many of her peers from mainland China, she came to science as a reward for being a top student. She obtained a bachelor's degree from the prestigious University of Science and Technology of China, and then enrolled at Harvard to complete a doctorate, focusing on the physics and chemistry of Earth's deep interior.

One of Jie Li's most creative studies focused on carbon in Earth's inner core.[66] Working with her graduate student Bin Chen (now a professor at the University of Hawaii) and a team of Deep Carbon Observatory colleagues, Li studied a superdense compound with iron and carbon atoms in a 7-to-3 ratio. Previous researchers had suggested that this unusual iron carbide might represent a promising deep Earth mineral, so the Michigan team put the idea to the test, squeezing the black, powdered sample between their diamonds to almost 2 million atmospheres while measuring a variety of physical properties. Extrapolating their results to inner-core conditions, they found an almost perfect match with the seismological observations—P waves similar to iron, but S waves much slower than pure iron. This finding is by no means proof that carbon in the form of iron carbide resides in Earth's inner core, but for now it seems like the hypothesis to beat.

In a complementary study published a few months later, a team of scientists led by doctoral student Clemens Prescher at the Bavarian Research Institute of Experimental Geochemistry and Geophysics (Bayerisches Geoinstitut, or BGI) in Germany subjected the same compound to simultaneous high pressure and high temperature and found unusual elastic properties, described as "rubbery."[67] That's not a typical description of mineral behavior, but it underscores how much we have to learn about carbon deep within our planet.

—

Our pursuit of the mysteries of Earth's core reveals a fundamental truth about science. We may set out to catalog all of the crystalline forms of carbon in Earth—the hundreds of known crustal minerals and scores of missing species, the dense carbonates of Earth's mantle, and the tantalizing hints of carbides in the core. But such a catalog, however complete, is not an end in itself. Our growing knowledge of the forms of Earth's carbon is leading us to an ever more vivid picture of our dynamic planetary home—how it emerged, how it works, its ultimate fate, and why it is unique in the Cosmos.

RECAPITULATION—Carbon Worlds

Earth's Mineralogy Is Unique[68]

What can carbon mineralogy tell us about our planetary home? Are we special? Earth certainly differs from the other rocky planets and moons in our own solar system. Mars, once a warm and wet world, has revealed only small and widely dispersed beds of presumed carbonates. Meteorites are similarly poor in carbon-bearing minerals and, in spite of meticulous scrutiny, the Moon has yielded only microscopic grains of graphite and iron carbide and not a single carbonate mineral. But what of more distant planets orbiting other stars?

One of the many useful outcomes of Grethe Hystad's mathematical explorations of mineral rarities was a ranking of all mineral species according to their probability of occurring on Earth. So, we asked, if we could start with another planet, identical to Earth in every way—the same size and mass, the same composition and structure, with oceans and atmospheres and plate tectonics—and replay 4.5 billion years of its history, and if we were lucky enough to discover 5,000 mineral species on that distant planet, how likely is it that they would be the same 5,000 minerals we see on Earth today?

I suspect that most mineralogists, when asked that question, would answer as I would, that the planet's mineralogy would be basically the same. All of the rock-forming minerals—quartz, feldspar, pyroxene, mica, and more—would be abundant, to be sure. Hundreds of other minerals, including diamond, gold, topaz, and turquoise, even if relatively scarce, would inevitably occur as well. And, by extension, I would have guessed that almost all of the rare minerals would also appear on any Earthlike world. They would still be rare, but they would be found eventually.

Not so, according to Hystad's calculations. Replay the tape and it's likely that roughly half of the species—more than 2,500 minerals—will be exactly the same, found on virtually all planets that have Earth's chemical and physical characteristics. Another 1,500 minerals, though not exactly common, also have a good chance, perhaps 25 to 50 percent, of appearing on both of any given pair of Earths. But compare any two planets and more than 1,000 of the rarest mineral species are likely to differ, with many minerals occurring on fewer than 10 percent of Earthlike worlds.

From these estimates, it was an easy step to calculate the chance that two planets would have identical mineralogies: just multiply together the individual probabilities for all 5,000 mineral species. We were stunned at the result. The odds against such a coincidence are literally astronomical—greater than 10^{320} (that's a 1 with 320 zeros)!

Compare that incomprehensible figure to estimates of the number of planets in the Cosmos. By the most favorable models of the known Universe—a hundred trillion galaxies, each averaging a hundred billion stars, and making the unlikely assumption that every star holds an Earthlike planet—there are at most 10^{25} planets like our own. Stretching these numbers to absurdity, you would need to examine every planet in almost 10^{300} universes like ours just to find one planet that closely replicated the minerals of Earth.

Hystad's striking conclusion was published in a 2015 issue of *Earth and Planetary Science Letters*: "In spite of deterministic physical, chemical, and biological factors that control most of our planet's mineral diversity, Earth's mineralogy is unique in the Cosmos."[69]

A deep philosophical aspect underlies Hystad's discovery—an implication related to an age-old debate regarding the relative roles of chance versus necessity. Complex systems, whether of minerals or of living organisms, evolve in both deterministic and stochastic ways. On the one hand, many aspects of nature are inevitable, dictated by the laws of physics and chemistry. Drop a pebble and it will fall; set fire to a piece of paper in Earth's oxygen-rich atmosphere and it will burn. On the other hand, all complex systems experience singular events—"frozen accidents"—that also define evolutionary pathways. The tension between chance and necessity is heightened because in most natural systems it's not always easy to tell which is which. Why does one rare mineral form but not another? Why does Earth have such a large Moon? Why did intelligent life emerge on Earth? Was it chance or was it necessity?

In mineralogy we can now, to a remarkable degree, resolve this tension quantitatively. Our conclusion is that, while many aspects of Earth's mineralogy are deterministic, chance also plays a fundamental role. Rare minerals form through unlikely sequences of chemical, physical, and biological processes. Consequently, Earth is absolutely, unambiguously unique in the Cosmos. Perhaps that's a good thing.

"Earthlike" Planets[70]

Few topics in science command as much attention as the discovery and description of extrasolar planets—those unseen worlds light-years beyond our own Sun. In humankind's quest to know whether we are alone in the Cosmos, astronomers are teasing out subtle wobbles and periodic dimmings of distant stars—telltale signs that a planet, much too faint to see directly in telescopes, is nevertheless in orbit.

First to be discovered were alien giants more massive than Jupiter, whizzing around their nearby stars in frenetic orbits of a few days, thus imposing the maximum possible stellar perturbations. But as we pass the twentieth anniversary of the first planet to be discovered beyond our own solar system, focus has shifted from behemoths to worlds more like Earth.

The term "Earthlike" evokes different meanings for different people. Astronomers focus on three characteristics that they can confidently measure: radius, mass, and orbit. Earthlike radii are inferred from the maximum dimming of the star as the planet eclipses a tiny fraction of its light, while mass derives from the extent of stellar wobbling induced by gravitational effects. In addition, for a planet to be given the "Earthlike" label, its orbital parameters must place it within the "habitable zone"—the flattened doughnut-shaped volume of space where liquid water might persist at or near the planet's surface. Growing numbers of discoveries—Kepler-186 f, Kepler-438 b, Kepler-452 b (all identified by data from the Kepler space telescope)—approximate these astronomical constraints. So do at least three of seven planets orbiting TRAPPIST-1, a small star just 40 light-years from the Sun. Almost monthly headlines herald the "most 'Earthlike' planet yet."

What those giddy articles usually fail to mention is that radius, mass, and orbit are, by themselves, rather poor indicators of Earth's potential planetary twins. What's missing is chemistry. Visible-light spectra of distant stars—data easily acquired with modest telescopes—reveal that stars differ widely in their chemistry. Some stars have a lot more or a lot less magnesium or iron or carbon than our Sun has. And it's likely that those critical compositional differences are mirrored to a significant extent in the makeup of their companion planets, because they are formed from the same protoplanetary disks.

Planetary compositions matter. Recent studies by mineralogists and geochemists suggest that even small differences in composition might render a planet ill suited to life. If there's too much magnesium, then plate tectonics, the engine essential to life's cycling of nutrients, can't get started. Too little iron, then a magnetic field that protects life from lethal cosmic rays can't form. Too little water or carbon or nitrogen or phosphorus, then life as we know it fails.

So, what are the chances of finding another Earth? With more than a dozen key chemical elements and scores of minor ones, the chance of replicating all the critical compositional parameters is small: perhaps only one in 100, or maybe one in 1,000, "Earthlike" planets will be compositionally similar to Earth. Nevertheless, with a conservative estimate of 10^{20} planets similar to Earth in radius, mass, and orbit, countless worlds must be rather like our own.

That realization should give us pause. It's only human to want to find planetary partners that remind us of Earth, just as we seek friends and lovers who share our tastes, our politics, and our religious beliefs. But to stumble across someone who is exactly like us in every respect—dresses the same, has the same profession and hobbies, uses exactly the same idiosyncratic phrases and body language—would be a little creepy. In the same way, I think we might find it a bit disturbing to discover a clone planet—one in many ways indistinguishable from Earth.

Not to worry; it isn't going to happen. So, as we boldly search for ever more "Earthlike" planets, we can be equally confident that there is only one planet truly like Earth.

CODA—Unanswered Questions

CARBON, THAT FLAMBOYANTLY versatile element of Earth, has taught us much about our world. We have cataloged hundreds of carbon-bearing minerals from tens of thousands of localities. We have learned to make thousands of synthetic analogs of these minerals, many with untapped commercial and technological applications. And we've even begun to predict what remains to be discovered, to glimpse the unknown in the crust and much, much deeper within our planet.

Yet all that we know about carbon in Earth pales compared to what we don't know. What new exotic minerals are waiting to be found, both in the crust and in much deeper realms? What are the structures and properties of those unknown compounds? How might they affect our lives? How much carbon is hidden deep within our planet, 99 percent of which may be sequestered in Earth's core? We scientists will continue our passionate quest for answers. Experiments, theory, and observations of the natural world will consume us for decades to come.

We are equally ignorant of how carbon moves from one reservoir to another, especially from the deep interior to the surface and back again. For that story, we must look to the skies—Earth's protective blanket of air—and to carbon, the element of cycles.

MOVEMENT II—AIR

Carbon, the Element of Cycles

Earth frames our lives. It is Earth on which we plant our crops,
build our homes, live our lives, and bury our dead.
Carbon, in its endlessly varied crystalline forms, in the richness
of its mineral kingdom, and in the discoveries of new materials,
epitomizes the essence of Earth.

Earth does not exist in isolation. Surrounding, embracing
solid Earth is the gentle, wispy, insubstantial sphere called Air.
Air emerges, the child of Earth. Volcanic wounds bleed
gases into the sky—a translucent blue blanket of protection.
Air shields us from the terrifying nothing of space,
keeping our planet temperate and sustaining life.
But human-induced changes may be altering Air
in ways as yet uncharted.

Wonder and worry—conflicting emotions we experience
when thinking of our planet's past, present, and future.

INTRODUCTION—Before Air

C ARBON MOVES, cycling from one sphere of Earth to another in an endless dance. Atmosphere, hydrosphere, biosphere, geosphere—all contain their share of Element 6, and all participate in the global carbon cycle. So it has been for more than 4.5 billion years, though the nature and extent of the cycle has changed through deep time and many details of carbon's movements remain hidden from view.

Today's Earth, our familiar white-and-blue marbled living world, bears little resemblance to its infant form. Earliest Earth, a blasted rocky landscape, lacked an enveloping cradle of air. Our home was born of dust—far-flung and tenuous, and hardly the stuff from which you'd think to make a respectable rocky planet.[1] But the Cosmos is endowed with prodigious quantities of dust, and dust tends to clump together (as you know if you occasionally clean those neglected recesses behind your dresser or under your bed). And so, as the proto-Sun ignited, subjecting the inner solar system to refining pulses of heat, primordial dust bunnies melted into the tiny droplets called "chondrules." As those sticky droplets clumped together, the first generation of space rocks was born. Grape-sized pebbles, fist-sized cobbles, and larger rocks the size of buses and buildings emerged from the dusty nebular milieu. Innumerable rocks orbited the faint young Sun, colliding and clumping into ever-larger incipient worlds.

Gravity held sway throughout the solar realm. Larger masses inex-

orably lured the smaller ones, capturing them in their gravitational
wells and swallowing them whole, sweeping clean broad doughnuts
of space. Orbiting 90 million miles from the still-growing Sun, the
proto-Earth emerged as the largest of the competing planetesimals.
What a spectacle that emerging Earth must have been! More than 4.5
billion years ago, numberless boulders rained down upon the growing world that would become our home. Falling faster than speeding bullets, the rocks in each collision delivered kinetic energy that
transformed to searing heat and blinding light upon impact, splash-
ing immense fountains of incandescent magma high above the fiery,
molten surface. Irregular chunks of iron-nickel metal, mountains of
silicate minerals, immense fluffy snowballs rich in water, and the occasional black, carbon-rich rock added to the growing, glowing sphere.

Epic physical processes with esoteric names—accretion and differentiation, fractionation and condensation, crystallization and
convection—transformed the nascent Earth from a random, stew-like
hodgepodge to a more rational, chemically ordered globe. Gravity
drove Earth's differentiation into a layered, structured world with the
densest components, principally a mix of molten iron and nickel metals, sinking to the center to form the core. Carbon may still play a
minor role in that deep and utterly inaccessible region, moderating
the extreme density difference between core and mantle, altering the
physical behavior of that hidden realm, but we are far from knowing the details of that story. We can be sure that carbon in the core,
if indeed it resides 3,000 miles or more beneath our feet, can play no
significant role in the air we breathe.

Surrounding the iron core, like the juicy fruit encircling the hard
pit of a peach, is Earth's thick, stony mantle. Minerals rich in the
lighter elements silicon, magnesium, and oxygen dominate the outer
2,000 miles of Earth. At the extreme conditions of the deepest mantle,
where pressures exceed a million atmospheres, mantle rocks are denser
than their near-surface counterparts. Still, those rocks float, buoyed by
the much denser liquid metal of the outer core, just as a pebble easily
floats on a pool of dense liquid mercury.

The disrupted, partially reflected passage of speeding sound waves, or "seismic waves," through solid rock reveals more hidden layers. Moving outward, the mantle features three broad, spherical zones. The all-but-inaccessible lower mantle extends from below about 400 miles to the core-mantle boundary almost 2,000 miles down and represents more than half of Earth's volume. The transition zone of intermediate density occupies a relatively narrow shell at a depth of about 300 to 400 miles, while the upper mantle extends almost to the surface.

Earth's outer layers—the crust, the oceans, and the atmosphere—are its thinnest zones, like the shell of an egg. Together they account for less than 100 miles of Earth's 4,000-mile radius and only about 1 percent of its mass. Nevertheless, these surface layers are by far the most chemically diverse, for they concentrate many of the rare elements that find no comfortable crystalline home in deeper minerals. These outer layers are also the most variable in depth. Ocean crust extends only 5 or 6 miles down in places, in contrast to the continental crust, which may exceed 50 miles in thickness beneath the tallest mountain ranges.

Directly exploring depths more than a few miles down is beyond any modern technology, but scientists have devised other means to understand Earth. We collect the rocks, scoop up the water, and sample the air at field sites on every continent, from the equator to the poles. From every corner of the globe the story is the same: rocks, water, air, and life have coevolved over billions of years of Earth history. And all of these diverse materials are critical rest areas in Earth's dynamic deep carbon cycle.

ARIOSO—The Origin of Earth's Atmosphere

T HE ATMOSPHERE WARMS EARTH. It shields us from the Sun's harshest radiation. It sustains us, providing the oxygen we breathe and the water we drink. The air also harbors vast stores of carbon that are consumed by the plants we eat. Yet Earth at its birth and in its earliest infancy had no atmosphere. Air had to emerge, to grow from nothing. How?

To find out, let's imagine ourselves back in time more than 4.5 billion years, to an epoch when planets were still forming, when the solar system was in chaos, and when Earth's carbon cycle was just getting started.

Earth Receives Carbon from Space

Four and a half billion years ago. The proto-Earth has coalesced. Its multilayered structure of core, mantle, and crust has begun to emerge. Dozens of chemical elements, most of them rare, have fallen to Earth in showers of stones.[2] A few mineral-forming ingredients dominate the mix—iron, silicon, magnesium, and, most abundant of them all, oxygen—contributing 90 percent of the mass. Calcium, aluminum, nickel, and sodium account for 90 percent of the rest. The remaining richness of the periodic table occurs in minor amounts—a few atoms per thousand of nitrogen and phosphorus, a few atoms per million of lithium and fluorine, and a few atoms per billion of beryllium and gold.

Earth and the other "terrestrial planets"—Mercury, Venus, and Mars—formed closest to the Sun, whose relentless heat boiled off most gases. Hence, these rocky inner planets favored elements that could form solid minerals. By contrast, the most abundant elements in the Cosmos—gaseous hydrogen and helium—were largely swept far away from the Sun, pushed out to distances of a half-billion miles or more by powerful solar winds to the domain of the "gas giant planets": Jupiter, Saturn, Uranus, and Neptune.

On a cosmic scale, hydrogen and helium are always the biggest part of the story, accounting for 99 percent of all atoms. Of the remaining 1 percent of atomic matter—the stuff that forms rocky planets—carbon atoms play a key role. Carbon represents almost one in every four atoms that is not hydrogen or helium. Only iron and oxygen are more abundant among the planet-forming leftovers. Yet, unlike iron and oxygen, most carbon atoms early in the history of the Universe were locked away in small volatile molecules of carbon dioxide, carbon monoxide, and methane. Earth's resulting inventory of carbon is meager, though poorly constrained; our best guess is no more than one atom in a hundred.

All of Earth's carbon came from space, originating in three major sources. A small portion came directly from the solar wind, enriched in the Sun's carbon-bearing gases. A greater fraction came to Earth as black, carbon-rich meteorites, which still fall from the sky from time to time.[3] These fascinating stones brought all manner of organic molecules—fuel-like hydrocarbons and alcohols, essential biomolecules including the amino acids, sugars, purines, and pyrimidines that play crucial roles in DNA and RNA—all prefabricated and ready to undergo chemical transformations. Most important of all for Earth's growing carbon inventory were comets, which are especially rich in small gaseous molecules like carbon monoxide and carbon dioxide, not to mention much of the water that would become Earth's encircling oceans.

Much of that carbon inventory circulated in fluids deep beneath the surface, there subjected to temperatures sufficient to break down large collections of atoms into their simplest molecular components— nitrogen, water, and carbon dioxide. And so Earth's carbon cycling began, for rocks cannot hold hot, pressurized, mobile fluids for long; they strive toward the surface, seeking any possible escape route. They hitch a ride dissolved in molten rock—red-hot magma that works its way upward, exploiting any crack or fissure. Close to the surface— perhaps a mile down, perhaps less, where the pressure drops below a critical value—the hot fluids transform into explosively expanding gas. Like champagne from an uncorked bottle, pulverized rock and gases blast outward, creating incandescent fountains of ash and air. In cooler zones, those buoyant, mobile molecules must have also risen through the crust, searching for any way to escape, and gradually seeping out of the vast expanse of passive land. Water thus released became the first oceans, while the gases became the first air.

No one knows the composition of Earth's earliest atmosphere.[4] The chemically inert gases that dominate today's air—primarily nitrogen in the form of N_2 molecules, with a bit of argon—were surely present from the start. Earth's atmosphere would have to wait more than 2 billion years for the life-induced buildup of reactive oxygen, O_2, its other major modern-day component. Smelly volcanic, sulfur-bearing gases such as hydrogen sulfide (H_2S) and sulfur dioxide (SO_2) must have been part of the primordial mix, while carbon-rich gases also contributed to Earth's earliest atmosphere.

The atmosphere's carbon is concentrated in three simple molecular species. Carbon dioxide, CO_2, gets the most press these days (most of it bad). It's a simple molecule—a central carbon atom flanked by oxygen atoms in a neat little row. At the cold temperatures of outer space, carbon dioxide can freeze into clear, colorless crystals known as "dry ice." On Earth, CO_2 is the dominant carbon-bearing species of the atmosphere—400 parts per million according to recent measurements, and rising by the year.

In deep space, far beyond any star or planet, where isolated atoms find each other only rarely, a single oxygen atom can bond to one carbon atom in carbon monoxide, CO—one of the most common molecules in space. Carbon monoxide has always been a minor component in Earth's atmosphere; it's less than 1 part per million of the air we breathe today. In our everyday lives, carbon monoxide poses a real and present danger because it is easily produced when carbon-based fuels are incompletely burned. Carbon fuels always combine with oxygen, but if the flow of air to your furnace or fireplace is obstructed and there's not enough oxygen to produce carbon dioxide, then carbon monoxide may flood your home. The consequences are devastating.

Carbon monoxide is insidious because it is colorless and odorless, and our bodies treat it just like oxygen, O_2. Yet, unlike oxygen, which we quickly consume in respiration, carbon monoxide blocks human respiration. Starved of oxygen, you slowly lose consciousness. The brain dies, and then you die.

The third simple carbon-bearing molecule in Earth's atmosphere is methane, CH_4—what you pay for as "natural gas" when you cook your food or heat your home. It's an elegant little molecule with one central carbon atom surrounded by a pyramid of four hydrogen atoms. Earth's modern atmosphere holds only trace amounts of methane—just 2 parts per million—but, as we will see, 2 parts per million is enough to have consequences.

Air—Starting Over

In a sense, Earth's most ancient atmospheric mix doesn't matter a whit, for in one shocking instant, almost 4.5 billion years ago our planet's protective blanket of air was obliterated.

The earliest history of Earth's atmosphere was filled with drama. Great volcanoes belched steam and air from the deep interior, even as volatile-rich comets rained down from space. As the thickness of the enveloping gaseous layer grew, the impacts of great stones from space pummeled the atmosphere, occasionally blasting off the outer reaches. Those speeding space rocks, some hundreds of miles in diameter, may have disrupted and mixed Earth's outer layers, but they could not stop the steady differentiation of rock, water, and air.

One collision was different—much larger and more destructive than any other event in Earth's history. For tens of millions of years, as the Sun's planets were forming and competing for orbital space, Earth had a worthy competitor—a rival world of planetary proportions. It's been called Theia, after the ancient goddess mother of the Moon. Possibly larger than Mars, but much smaller than the still-growing Earth, Theia competed for the same orbital real estate as Earth. For a time, perhaps for a few tens of millions of years, Earth and Theia played their dangerous dance at a distance. Near misses may have perturbed their twin orbits, setting into motion what would become their inevitable final duel.[5]

Laws of gravitation dictate the evolution of solar systems; it's a rule that two planets can never share the same orbital space. At some point they will come too close, and when they do, be sure to bet on the larger world.

One dramatic day, when Earth was still in its infancy—perhaps a mere 50 million years old—Theia smacked into Earth. Some models of the event suggest a sideswipe, an almost glancing blow, yet one that was fatal to Theia. The result was utter obliteration: a space-bound observer of the planetary spectacle, watching from a presumably safe distance, would have been mesmerized to see Theia ripped asunder, largely vaporized in white-hot death throes.

Observing such a catastrophe, you could be forgiven for failing to notice that another victim of the impact was Earth's tenuous atmosphere. Every molecule of air was displaced, mostly ejected into deep space, never to return to what had quickly become the uncontested third planet from the Sun. More noticeable was the immense glowing cloud of vaporized rocks—a mix of incandescent debris from the mantles of both Theia and Earth. A significant fraction of the mass fell back down to Earth's encircling magma ocean as torrential rains of red-hot molten drops. Another fraction was flung into orbit, soon to gather itself into Earth's companion Moon.

It was time to start again. The great reset button that was the Moon-forming event also began the formation of Earth's atmosphere anew, as the deep carbon cycle began in earnest.

Clues about Earth's Earliest Atmosphere

More than 4 billion years ago, the atmosphere was a work in progress. Geologists aptly name Earth's unsettled first half-billion years or so the "Hadean Eon." Were you to have ventured onto that newly resurfaced Earth, your overwhelming impression would be of an unrelentingly violent and hostile place. It's difficult to say which was more to be feared—the incessant rain of rocks from the sky, or the ferocious explosive volcanism from below. But those twin dangers together delivered carbon to the air and set Earth's carbon cycle in motion. A new atmosphere literally rained from the heavens and erupted from the depths.

How can we possibly guess the nature of the ephemeral atmosphere that enveloped Earth more than 4 billion years ago? Precious few microscopic mineral grains survive from those earliest of times, while only a smattering of rocks from more than 3.5 billion years ago persist in remote terrains of Australia, Canada, Greenland, and South Africa. Those scattered bits are all but mute when it comes to the nature of ancient air.

Nevertheless, clues to ancient air exist. Three lines of evidence from three different scientific domains—astrophysics, geochemistry, and planetary science—hint at what might have been.

Clue #1— The Faint Young Sun[6]

The first revealing clue about Earth's Hadean atmosphere comes from what might seem an unlikely source: physicists who study the evolution of stars. Stellar astrophysicists tell us that the Sun is in the midst of a multibillion-year period of stability, a time enjoyed by most stars as they steadily consume hydrogen in nuclear fusion reactions to make helium. Look up into the night sky. Nine of every ten stars you see are "burning" hydrogen in nuclear fusion reactions, as a fraction of a percent of their hydrogen mass is converted to heat and light. These reactions are what we see and feel as sunlight.

But there's a catch. Hydrogen-burning stars change slowly, steadily brightening over time. That change is not enough to be a noticeable difference from one century to the next, or even over the span of a million years, yet over billions of years the Sun has become much more luminous. More than 4 billion years ago, the Sun radiated at only 70 percent of today's output. That is a huge difference; such a drop in the Sun's radiant energy, if experienced today, would have immediate and catastrophic effects. Earth would freeze, with ice extending from the poles to the equator. Life would almost cease to exist, with small, localized colonies of simple organisms clinging to warm, wet zones near volcanic vents.

As we think back 4 billion years to the time of that faint Sun, we must wonder, How did Earth avoid becoming encased in ice? The few mineral and rock scraps that survive from Earth's first half-billion years certainly don't point to a frozen world.

Greenhouse gases appear to be the most plausible solution. Much as a gardener's greenhouse can stay warm even on cold winter days, some atmospheric gases possess the ability to absorb and trap the Sun's energy, reducing the amount of heat radiating back out to cold space. Water vapor and clouds have always been part of the greenhouse equation; on Earth today, they account for about half of the vital modern greenhouse effect that prevents our planet from freezing over. But water molecules alone are not enough to have compensated for the faint young Sun. Earth needed other molecules to trap enough heat—molecules containing carbon.

Clue #2—Geochemistry

If large inventories of greenhouse gases prevented Earth from becoming a frozen planet more than 4 billion years ago, where are those gases now? Geochemists who conduct global inventories of Earth's elements point to the vast deposits of carbonate minerals found today on every continent—minerals that would not have formed in such abundance 4 billion years ago. A balance has long existed between carbonate minerals and atmospheric CO_2; every molecule of carbonate in the crust means one less molecule of carbon dioxide in the air. The conclusion is this: Four billion years ago, when there were fewer carbonate minerals, most carbon was locked in CO_2 in the atmosphere, with an air pressure perhaps several times the present value.

Some geochemists posit a wrinkle in this scenario; they think the carbon dioxide–rich atmosphere was amplified by methane, a gas that might have been much more abundant before the rise of oxygen 2.5 billion years ago. Methane is a potent greenhouse gas, molecule for molecule many times more effective than CO_2. Abundant methane would have been bathed by cosmic rays, triggering varied organic chemical reactions and producing a haze of molecules that may have given Earth's earliest skies a distinctive orange tint, just as is observed on Saturn's moon Titan today.

Were today's Earth suddenly surrounded by such a thick atmospheric mix of carbon dioxide and methane, the climate would swing wildly to unprecedented hothouse conditions. It's a question of balance. The greenhouse effect is essential to life; today's Earth would freeze over from poles to equator without it. But too much greenhouse gas means that too much heat becomes trapped. We can reach an atmospheric tipping point, when warming releases more methane and carbon dioxide from soils and rocks, which in turn leads to more warming—a positive feedback, perhaps leading to an irreversible runaway greenhouse effect.

What would happen if all of the carbonate minerals in Earth's crust were transformed into atmospheric carbon dioxide? What would happen if that prodigious reservoir of more than 200 million billion tons of carbon—more than 100,000 times the concentration of the modern atmosphere—were suddenly converted to a gas? The answer lies in clear view: Earth would become like Venus.[7] Venus is, in many respects, Earth's planetary twin—the same size, same density, and same basic composition. But the combination of its orbit 25 million miles closer to the blazing Sun and a dense CO_2 atmosphere at a pressure 90 times that at Earth's surface has led to a runaway greenhouse effect. Surface temperatures on Venus average 900 degrees Fahrenheit—hot enough to melt lead.

Perhaps Earth was just lucky. (We have been called a "Goldilocks planet.") If so, carbon was a principal reason why.

Clue #3—Meteorites from Earth[8]

A third, much more speculative line of evidence—meteorites from ancient Earth—might reveal exquisite details of Earth's earliest atmosphere. This idea isn't as crazy as it might sound. More than a hundred meteorites have been positively identified as coming from Mars, because rocks are blasted off the red planet's surface when large comet or asteroid impacts disrupt the landscape. The smoking gun proving that these rather nondescript rocks come from Mars, as opposed to an asteroid or some other body, is the idiosyncratic combination of gas molecules preserved in minuscule air pockets. That mixture exactly matches the ratios of gases measured by NASA probes in the Martian atmosphere.

So, imagine the aftermath of one of the giant impacts of an asteroid on Earth 4-plus billion years ago. Chunks of excavated rocks must have been hurled out into space. Those rocks must have incorporated tiny bubbles of Earth's ancient atmosphere. Those bubbles must still be entombed in protective minerals. So all we need to do is go to the Moon and find a few of the countless thousands of Earth meteorites that must have fallen onto the surface of our nearby luminous companion. Indeed, many of us think the recovery of Earth meteorites is one of the most compelling reasons to go back and walk again on our nearest celestial neighbor.

To collect a tiny bit of Earth's earliest air—now that would be something!

INTERMEZZO—The Deep Carbon Cycle

IKING IN THE BUCOLIC HILLS of the Caldara di Manziana in central Italy, surrounded by forests, flowers, and singing birds, you don't expect to stumble across bold signs emblazoned with a skull and crossbones, warning of death.[9] What could it be? An electrified fence? A firing range? Errant bears?

And then you come upon the little valley—the lifeless depression, a swath of bare soil in stark contrast to the verdant higher ground. What's going on?

The culprit is carbon dioxide. It seeps from the ground, an invisible, odorless gas. Heavier than normal air, CO_2 hugs the ground, filling the lowest depressions. On a breezy day it doesn't really matter. The flow of subterranean gas is quickly, harmlessly dispersed. But on a calm day with no wind, the concentration of denser carbon dioxide will displace breathable air, creating a deadly scenario. Hunters have been the victims. Their dogs, lower to the ground, collapse and die first. If a hunter rushes to the aid of his trusty companion and kneels beside the stricken animal, he, too, may be overcome, oblivious to the deadly danger.

—

Carbon moves. As an ingredient of Earth's transient ocean crust, it plunges from the sunlit surface to the deep interior through subduction. As an essential component of deep mantle fluids, it seeps out of the ground from soils and bursts forth from explosive volcanoes. Carbon atoms precipitate as solid rock from oceans and air, and they weather away from solid rock to return to the oceans and air. Once released, carbon atoms flow around Earth in stately ocean currents and drift across the globe in the changeable atmosphere. All the while, living cells from microbes to plants to people use and reuse carbon atoms at a pace that far outstrips the cycling of Element 6 on any known nonliving world. Rare, indeed, is the carbon atom that remains locked away for more than a tiny fraction of Earth's 4.5-billion-year history.

Italy's CO_2-rich volcanic exhalations—the ones that occasionally kill dogs and their masters—arise when incandescent carbon-bearing magmas ascend from the depths. Some of those magmas interact with buried layers of carbonate minerals, which decompose under the intense heat, mingling mantle and crustal sources of carbon dioxide. It's all part of a grand cycle of carbon in Earth—the cycle that created and replenishes the air.

All of Earth's chemical elements experience cycles, and carbon is no exception. The carbon cycle, a mainstay of introductory textbooks and popular science websites, consists of varied reservoirs of carbon atoms, as well as movements of those carbon atoms among the reservoirs.[10] Search for "carbon cycle images" on YouTube and you'll find oceans and atmosphere, limestone and fossil fuels, animals and plants, all with little arrows showing how the carbon moves from one reservoir to another. Some of these diagrams also add a smoking volcano, hinting at deeper processes, but Earth's deep carbon—the ultimate atmospheric source—is rarely considered in any detail.

The reasons for this neglect are easy to understand. Relative to the rapid carbon turnover near Earth's surface, the deep carbon cycle is slow. It takes many millions of years for a carbon atom to journey deep into the planet and return to the surface, and details of the processes remain largely hidden and uncertain. No one knows how much carbon is down there, nor are we sure of the varied forms it takes.

What we do know is the grand global mechanism by which carbon cycles from Air deep into Earth and back again. The black basalt and other rocks that pave the floors of Earth's oceans are cold and dense— denser than the hot, soft mantle beneath. Gravity wins, as immense slabs of ocean crust "subduct," diving down hundreds of miles, carrying with them layers of sediment and basalt rich in carbonate minerals and the decomposing detritus of life. Inexorably, carbon is transported from the surface, plunging deeper and deeper into Earth's inaccessible interior.

If the removal of surface carbon continued in this way unabated, with no recharging of surface carbon reservoirs, then the crust could be stripped clean of carbon in a few hundred million years. Subjected to such deprivation, the carbon-dependent biosphere would collapse. Fortunately for the biosphere, what goes down also comes up. As the carbon-rich subducting rocks heat up, carbonate minerals and organic molecules start to break down to produce carbon dioxide and other small molecules. Some of those molecules break free of their rocky tombs to form buoyant fluids that reverse course and rise to Earth's surface. Erupting volcanoes are the most obvious points of release for these deep gases, but widespread diffuse seepage of deep carbon out of the ground and into the air may represent an even larger flux, albeit one that is difficult to quantify and correspondingly uncertain.

Understanding this global-scale carbon cycle, the one mostly hidden from view, has been one of the central goals of the Deep Carbon Observatory from its inception.[11] The work is rich and varied, with hundreds of scientists tackling diverse and challenging problems at scores of field stations and laboratories around the world. Their research on the dynamic deep carbon cycle can be summarized in three questions: What goes down? What happens to carbon when it's down there? What comes back up?

What Goes Down?[12]

The vast majority of the carbon atoms on Earth, more than 99.9 percent, are buried beneath the surface and locked away for millions of years in the crust and mantle. Most carbon atoms are stored in immense layered deposits of limestone, others in the buried biomass of coal or petroleum or other black, carbon-rich sediments. And a dynamic fraction of that hidden carbon is trapped in slabs of ocean floor basalt and sediments that plunge deep into Earth's mantle at subduction zones.

How do carbon atoms from Earth's changeable sunlit surface, those once mobile atoms in the air or the oceans or living cells, find themselves sequestered in solid rock, moving inexorably into the depths? It's a matter of chemical reactions, both abiotic and life driven. On the abiotic side, carbon dioxide in the atmosphere and oceans readily reacts with calcium and magnesium atoms in newly exposed volcanic basalt and other rocks to make carbonate minerals. These weathering reactions are usually hidden from view deep in the ocean or buried in soils, but occasionally, in locations where surface waters are rich in calcium or magnesium, you can watch carbonate crystals grow in real time as carbon dioxide is greedily pulled from the air.

Life learned the trick of making carbonate minerals hundreds of millions of years ago. For a time, carbonate "biomineralization" occurred exclusively in shallow waters along coastlines, near reliable supplies of mineral nutrients. Massive reefs—alive with corals, mollusks, and other armored animals—extended for hundreds of miles and locked away immense quantities of carbon. Other carbon atoms were sequestered in the form of biomolecules and their decay products, as cells died and were buried on land or at sea.

The history of life, well documented by fossils and the sediments in which they are entombed, saw repeated dramatic changes in the nature and extent of carbon burial. The rise of photosynthetic algae 2.5 billion years ago initiated what was probably the first major pulse of biomass in sediments, as algal mats bloomed in shallow, sunlit waters, died, and sank to the ocean floor. The invention of carbonate shells more than a half-billion years ago added to the inventory of carbon buried at sea.

Life ventured onto land more than 400 million years ago, adding new twists to the subsurface carbon cycle. Trees and other plants sequester lots of carbon; when buried, they form thick deposits of carbon-rich peat and coal. Root systems, themselves an integral part of a deep, hidden biosphere, break down rocks into clay minerals. Sheet-like clay particles, nanoscale bits too small to be seen even in the most powerful light microscopes, develop surface electrostatic charges that attract and bind biomolecules. Clay surfaces become festooned with a carbon-rich coating. Those carbon-bearing clay minerals erode from soils, flow down streams and rivers, and eventually arrive at the oceans to form thick sedimentary deltas. Earth's biosphere, it seems, has found many ways to bury carbon.

—

Two hundred million years ago, around the time when dinosaurs were beginning to dominate the land, microscopic organisms introduced another wrinkle to the carbon sequestration story.[13] That's when free-floating cells in the middle of Earth's oceans evolved the ability to make their own tiny armored carbonate plates by pulling carbon dioxide and calcium from their surroundings. These clever, carbonate-coated plankton, which live and die in prodigious numbers, were a game changer. For the first time in Earth history, carbonates formed near the sunlit surface of the deepest oceans, not just in reefs along the coastline. As dead cells sank to the ocean floor, they added a new source of buried carbon in sediments that had previously been devoid of carbonate minerals.

The implications of this "mid-Mesozoic revolution," as geologist Andy Ridgwell of the University of Bristol calls it, are profound. Prior to 200 million years ago, when life produced carbonates exclusively on the shallow continental shelves, the extent of limestone reefs waxed and waned with the highly variable level of the seas. During warmer periods, when global ice was minimal and sea levels relatively high, extensive reefs could form on the broad, submerged coastal plains. This enhanced growth of carbonate corals and shells moderated the concentration of calcium in the oceans—a key factor in determining seawater's variable acidity.

Contrast that scenario with cooler periods of extensive polar ice and glaciation, when sea levels were unusually low and continental shelves were mostly exposed to air. Few reefs could form. Lacking a biological path to make limestone, calcium in seawater rose to supersaturated concentrations, shifting ocean chemistry in the process. Ridgwell concludes that carbonate-forming plankton, thriving as they do whether sea level is high or low, have moderated ocean chemistry for the past 200 million years.

—

Prodigious quantities of Earth's carbon are extracted from air and water, gradually to be buried in the crust. Volcanic rocks react with carbon dioxide to form carbonate minerals. Biogenic carbonates form reefs at continental margins, and they rain down from the mid-oceans. Biomass is buried on land and at sea, while clay minerals absorb fragmented biomolecules and add more carbon to the thickening sedimentary inventory. Those carbon fluxes are easy to observe and to quantify.

Less obvious is the fraction of that buried carbon that continues its downward journey—the fraction that plunges deep into the mantle via subduction. The effort to quantify the changing input of subducting carbon has become a passion of Terry Plank at Columbia University's Lamont-Doherty Earth Observatory.[14]

From her laid-back style, you might not guess about Plank's adventuresome career or MacArthur "genius" status. She focuses on volcanoes as windows to the deep carbon cycle. In an effort to understand what comes out of erupting volcanoes, Plank decided to catalog the carbon and other elements in the ocean crust—the material that gets subducted down into the mantle. Comparing trace elements in those downward slabs with erupting magmas, she was able to show, in stunning detail, that recycled crust is a major component of many volcanic systems.

How much carbon makes it deeper into the mantle remains a matter of debate. Plank thinks the answer depends critically on where you look. "Some subduction zones will receive lots of carbonate, some none; some lots of organic carbon, some none," she observes. She also concludes that transporting carbon to depth can be difficult. Carbonate and biomass are less dense than basalt, and they tend to concentrate in the upper portions of the subducting slab—the part less likely to make it very far into the deep, dark interior. She concludes, "Carbon sub-duction requires some accidents."

In short, "what goes down" includes a lot of carbon buried in the crust in the form of biomass and carbonate minerals, much of which seems to come right back up. But the most intriguing and mysterious part of the carbon cycle is the fraction that takes the long, deep journey into Earth's mantle.

What Happens to Carbon When It's Down There?

As wet, subducting slabs carry carbon-rich minerals and black biomass to greater and greater depths, the temperature rises higher and higher. Biomolecules break down into smaller bits—mostly carbon dioxide or methane. Carbonate minerals also decompose, releasing more carbon-bearing molecules into a hot, water-rich fluid. Deep carbon is never alone. It always mingles with oxygen and hydrogen, usually with a bit of sodium, chlorine, sulfur, and other elements mixed in.

There's the rub. Understanding what happens to carbon in the mantle depends primarily on understanding water at high temperature and pressure. But a decade ago, when the Deep Carbon Observatory began, water in the mantle was *terra incognita*. No one knew in any detail the properties of H_2O at the extreme conditions of temperature and pressure hundreds of miles down.

The key unknown—the single parameter blocking our path forward—was water's dielectric constant, a measure of its "polar-ity." Water molecules assume a little V shape. A central oxygen atom links to two hydrogen atoms arranged like Mickey Mouse's ears. The hydrogen side of the molecule carries a positive electrical charge, while the oxygen side is negatively charged, yielding a "polar" molecule.

Many of water's most distinctive properties—its ability to dissolve table salt and numerous other chemicals, the ease of forming raindrops, the hardness of ice, capillary action in the stems of plants, and much more—arise from this polarity. The dielectric constant is a measure of the strength of that positive-negative charge separation, which dictates water's behavior.

We knew that water's dielectric constant changes dramatically with temperature and pressure, but at the start of the DCO adventure, we didn't know by how much it changed. Without that information, there was no way to calculate critical aspects of deep fluids—the solubility of salt, the electrical charges on dissolved molecules, or the acidity of solutions, for example. There was no way to predict the behavior of carbon or any other dissolved element in Earth's mantle. And so, at the May 2008 workshop that launched the DCO, Dimitri Sverjensky, professor of geochemistry at Johns Hopkins University, made a public plea to fill this gap in our knowledge.[15]

Sverjensky's talk lasted a mere five minutes, but it was effective. Shortly thereafter, Isabelle Daniel, professor of geochemistry at the historic University of Lyon in central France, sat next to Sverjensky at lunch. She, too, had been thinking about properties of mantle water, and she shared brand-new data on the behavior of carbonate minerals in water at extreme temperature and pressure—data that hinted at water's dielectric constant. Together, Sverjensky and Daniel hatched a compelling research plan that convinced the DCO to focus some of its resources on deep water. A decade later, their initiative has led to a revolution in our understanding of deep carbon.

Deep Water

Determining the dielectric constant of water at the high pressures and temperatures of Earth's mantle is a daunting problem, requiring advances on both theoretical and experimental fronts. Work in 2012 led by Giulia Galli and her graduate student Ding Pan at the Davis campus of the University of California tackled the theory part.[16]

They employed a quantum mechanical model to calculate that water's dielectric constant increases up to 100,000 atmospheres, to the point that carbonate minerals stable in Earth's crust begin to dissolve in the mantle. The profound implication is that carbon dissolved in water could be a major factor in the deep carbon cycle.

Meanwhile, Isabelle Daniel and her team in Lyon rose to the experimental challenge of testing Galli's predictions. Employing a sophisticated heated diamond-anvil cell, they measured how carbonate minerals dissolve under mantle conditions.[17] Experiments and theory yielded satisfyingly similar results, pointing to previously unsuspected complexities in the behavior of deep carbon.

Deep Earth Water Model

Quantifying the dielectric constant of water at mantle conditions was only the beginning in understanding the hidden deep carbon cycle. The new estimates of water's dielectric constant had to be incorporated into a comprehensive model of fluids at high temperatures and pressure. That "Deep Earth Water" model, or DEW model, as it has come to be known, was the inspiring invention of Dimitri Sverjensky.[18]

Few individuals have had such a profound impact on my scientific career. Sverjensky and I first met more than two decades ago on the Johns Hopkins campus at his serene, windowed office overlooking a wooded stream. I sought his help to understand the complexities of biomolecules interacting on mineral surfaces—a difficult problem that might be critical to understanding life's origins. Interested, but cautious, he explained that his inventive theory of mineral surface interactions worked well for individual ball-like metal atoms, but had not yet advanced to dealing with the much trickier problem of molecules, with their complex three-dimensional shapes.

Fast-forward to 2006, and Sverjensky was the one who contacted me. He had figured out the molecular adsorption problem, he had a yearlong sabbatical coming up, and he wanted to spend it at the Geophysical Laboratory. I was elated. We built a mineral surfaces lab, and for the next decade we enjoyed a steady stream of great students, dozens of research papers, and ample government funding. And in the process, we both became deeply involved with the DCO.[19]

Dimitri Sverjensky and I quickly became fast friends—a bond forged in part by a mutual love of music. He was born and raised in Sydney, the son of "The Great Sverjensky," the most revered piano teacher in all of Australia. Dimitri is an extraordinary scientist. At heart a modest and soft-spoken scholar, he is brilliantly creative, meticulous and rigorous, reserved, and averse to hyperbole. Yet it is not hyperbole to say that his DEW model shines as one of the crowning achievements of the DCO—an advance that will pay dividends long after the program's end.[20]

DEW-based discoveries, coming at an ever more rapid pace, reveal astonishing insights about Earth's deep carbon cycle. Prior models assumed that deep fluids are simple mixtures of water and carbon dioxide, but Sverjensky realized that atoms rearrange and form new types of dissolved molecules in a complex subterranean soup. Many of these molecules are "ionic species," with positive or negative electrical charges; consequently, carbonate minerals dissolve like salt, mobilizing carbon at depth.

Sverjensky also demonstrated beyond a doubt that Earth's mantle is a factory for the synthesis of organic molecules—the kind of carbon-based molecules that must have played a role in life's origins.[21] Under some deep Earth conditions, acetic acid (the principal component of vinegar) forms, but if the temperature or acidity shifts, then natural gas and other hydrocarbons will dominate the mix. Recent experiments by Isabelle Daniel reveal that much larger carbon-bearing molecules form too, including economically valuable ingredients of petroleum. All of these results point to a rich and complex deep organic chemistry that is only now coming to light.

Dimitri Sverjensky's most jaw-dropping finding relates to the origins of diamonds, which were universally thought to form as a consequence of carbon atoms being subjected to extreme pressure—a process in which water plays no significant role.[22] Working with grad-

Anatomy of Earth Depth under the surface

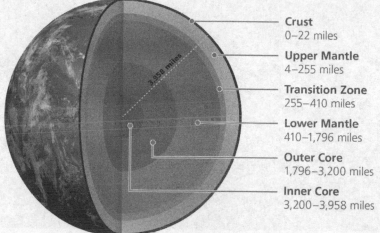

Crust	0–22 miles
Upper Mantle	4–255 miles
Transition Zone	255–410 miles
Lower Mantle	410–1,796 miles
Outer Core	1,796–3,200 miles
Inner Core	3,200–3,958 miles

The anatomy of Earth reveals several concentric layers, each at greater temperature and pressure than the one above. Carbon occurs in every layer of Earth, from its atmosphere to its core. *Adapted from a Deep Carbon Observatory image.*

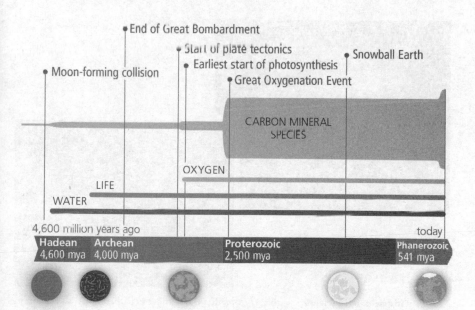

Earth's evolutionary time line spans more than 4.5 billion years. Life as we know it has evolved slowly and in concert with the geosphere over millions of years. Carbon has played a key role throughout this evolutionary path. *Credit: Deep Carbon Observatory/Josh Wood.*

Scottish scientist James Hutton realized that every aspect of the Siccar Point cliffs (pictured here) could be explained by the gradual natural processes that are at play all around us all the time. New sediments are deposited ever so slowly in layers and are gradually buried, heated, squeezed, and turned to stone, while older rocks are gradually deformed, uplifted, and eroded away—a net subtraction of rock layers. This layering is easily seen at Siccar Point. *Credit: Dave Sousa.*

Using a statistical model, scientists were able to predict the number of carbon minerals remaining to be discovered on Earth. The research gave rise to the Carbon Mineral Challenge, a citizen science effort to find Earth's missing carbon minerals. The first new carbon mineral, abellaite (left), was discovered in 2016; triazolite (right) was found a year later. *Photographs by Josep Soldevilla (left) and Marko Burkhardt (right).*

Professor William "Bill" Bassett did pioneering work in describing the first of a sequence of dense, high-pressure forms of calcite. He was an invaluable mentor to me, helping to define a good thesis project for a fledgling graduate student. *Photo courtesy of William Bassett.*

Diamond crystals, composed of pure carbon, are prized as gemstones. For research scientists, diamond inclusions, which were trapped and hermetically sealed as the growing diamonds engulfed them, reveal much about Earth's deep interior. *Credit: Stephen Richardson, University of Cape Town.*

A diamond-anvil cell, shown here being used by University of Michigan scientist Jie "Jackie" Li, enables scientists to create Earth's extreme range of temperature and pressure in the laboratory. The DAC makes it possible to synthesize materials not observed under surface conditions. *Photo by Allison Pease.*

Volcanoes, which regularly emit carbon dioxide and other gases into the atmosphere, are an important part of the global carbon cycle. Here, Deep Carbon Observatory scientist Tobias Fischer monitors gas emissions at the Poás Volcano in Costa Rica, in an effort to quantify volcanic contributions to the global cycle and potentially predict eruptions. *Credit: Carlos Ramírez Umaña.*

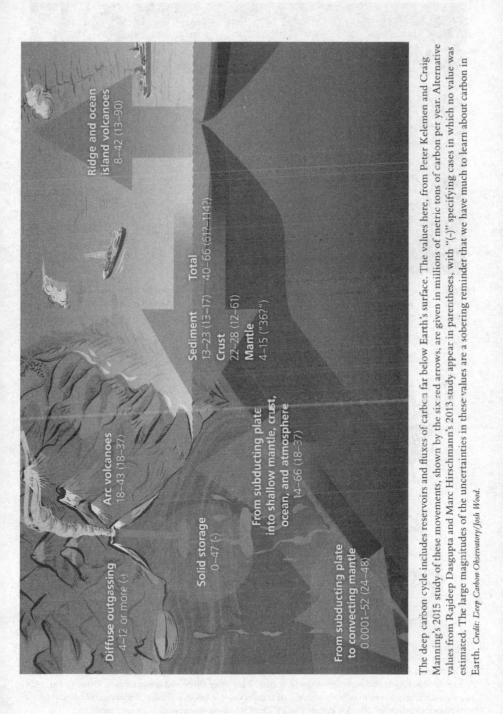

The deep carbon cycle includes reservoirs and fluxes of carbon far below Earth's surface. The values here, from Peter Kelemen and Craig Manning's 2015 study of these movements, shown by the six red arrows, are given in millions of metric tons of carbon per year. Alternative values from Rajdeep Dasgupta and Marc Hirschmann's 2013 study appear in parentheses, with "(-)" specifying cases in which no value was estimated. The large magnitudes of the uncertainties in these values are a sobering reminder that we have much to learn about carbon in Earth. *Credit: Deep Carbon Observatory/Josh Wood.*

Drilling projects on land and at sea reveal a rich, hidden, deep microbial biosphere. Here, technicians Margaret Hastedt, John Beck, Chad Broyles, Zenon Mateo, and Lisa Crowder carry a fresh core for processing. *Credit: Carlos A. Alvarez Zarikian.*

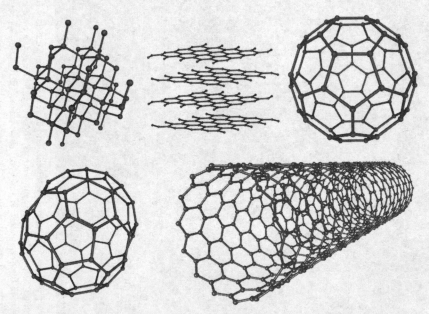

Carbon comes in many shapes and sizes, including (clockwise from upper left) the network structure of diamond, the layered structure of graphite, the soccer ball–like sixty-atom "buckyball," carbon nanotubes, and elongate enclosed forms. The latter three and other elegant forms of carbon are collectively known as "fullerenes," in tribute to the geometrically similar geodesic domes of American architect Buckminster "Bucky" Fuller. *Adapted from an image by Michael Ströck.*

Karen Lloyd and Donato Giovannelli, early career scientists at the Deep Carbon Observatory, investigate primitive microbes in Costa Rica. Biologists face myriad questions regarding when, where, and how life began on Earth. *Credit: Tom Owens.*

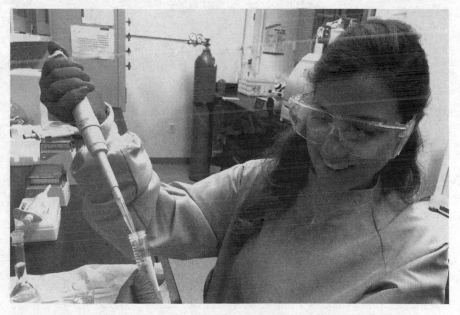

Teresa Fornaro studies interactions of carbon-based molecules with mineral surfaces at the Carnegie Institution's Geophysical Laboratory. Her work reveals possible steps in the origins of life. *Photo by Robert M. Hazen.*

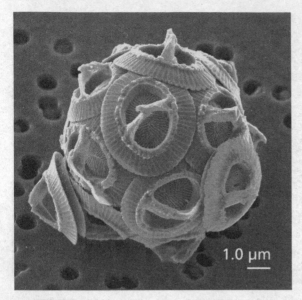

Coccoliths are microscopic carbonate plates that surround single-celled marine organisms. *Adapted by Deep Carbon Observatory/Josh Wood from a Wikipedia Commons image by NEON.*

1.0 μm

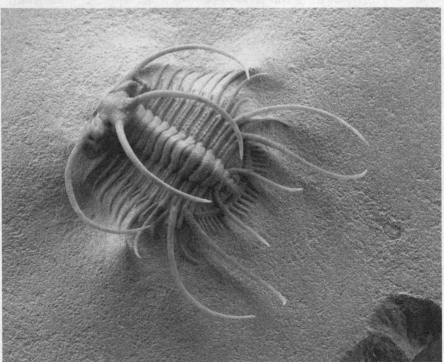

The trilobite *Boedaspis* from the Ordovician Period of Saint Petersburg, Russia, features a dramatically sculpted exoskeleton with numerous sharp carbonate spines. This specimen from the Hazen Collection is on view at the Smithsonian Institution's National Museum of Natural History. *By permission of Robert M. Hazen.*

uate student Fang Huang, Sverjensky found that diamonds form just as easily in watery mantle fluids without any change in pressure at all. Instead, simply increasing the acidity of the pressurized water-rich solution causes diamond crystals to appear. Indeed, natural fluctuations in deep water acidity could lead to cycles of diamond growth and dissolution—processes that match previously unexplained growth patterns of natural gemstones.

All of these discoveries and more hint at a dynamic realm 100 miles beneath our feet—a hidden domain of chemical invention that has played critical roles in Earth's deep carbon cycle for billions of years. But how can we know for sure? What evidence at Earth's surface might bolster Sverjensky's bold claims?

The Deep Methane Mystery[23]

Among the most provocative and far-reaching DEW findings is the demonstration that methane, potentially in prodigious quantities, may rise from Earth's mantle to form immense reservoirs in the crust. Geologists in other parts of the world, notably Russia and the Ukraine, have long advocated a deep abiotic origin for most natural gas and other hydrocarbons. Yet, many petroleum geologists in the United States and other oil-producing regions have argued vehemently against this idea; they point to large reservoirs unambiguously derived from dead plants, animals, and microbes. Some of us have long suspected that this debate, possibly exacerbated by Cold War animosities and professional rivalries, is a false dichotomy. Perhaps both camps are correct, and methane can form by multiple pathways. That is a possibility that DCO hoped to explore.

How do we test these competing hypotheses of methane formation—deep abiotic versus shallow biotic? What fraction of methane, if any, comes from hot mantle chemistry as opposed to microbial activity of the much cooler crust? How can we tell the difference? Aren't all methane molecules exactly the same?

All methane molecules are indeed CH_4, but it turns out C and

H come in a couple of different flavors—different isotopes that make methane studies a lot more interesting. Recall that carbon, which always has exactly 6 protons (the one characteristic that distinguishes carbon from every other element), can have 6 or 7 neutrons to make the more abundant carbon-12 (also shown as ^{12}C) or the slightly heavier carbon-13 (^{13}C). Hydrogen, which is the periodic table's first element, always with 1 proton, usually lacks any neutrons. However, about one in every 6,420 hydrogen atoms in ocean water has one neutron in the heavier isotope deuterium, designated D.

About 99 of every 100 methane molecules incorporate one carbon-12 atom surrounded by four normal hydrogen atoms. That's the most common natural form of methane. About one molecule in 100 has a heavier carbon-13, while only one in 1,500 holds a single deuterium. Things get really interesting when two heavy isotopes substitute, yielding either $^{12}CH_2D_2$ or $^{13}CH_3D$. Those rare "doubly substituted" methane molecules are literally one in a million. So, if you've been counting, that's five isotopically distinct kinds of methane to play with.

The exciting news for DCO scientists is that the ratios of these different isotope combinations have the potential to reveal methane's origins. Temperature plays a big role in the ratios, as does the sample's biological history. Theorists suggest, for example, that a methane sample formed at higher temperature will host slightly more $^{12}CH_2D_2$, while methane generated by microbes tends to be relatively depleted in $^{13}CH_4$.[24] All we have to do is measure the ratios of the five different kinds of methane in any sample to learn a lot about its origins.

Molecular mass is the key to unlocking methane's secrets. Each of the five varieties of methane has a slightly different mass. In theory, we could measure the masses of millions of individual methane molecules to determine their relative amounts and thus pinpoint their origins. In practice, it's not so simple.

Many laboratories routinely determine the abundances of $^{12}CH_4$ and $^{13}CH_4$, which differ by about 6 percent in mass. That's a relatively easy experiment. But when we began the Deep Carbon Observatory in 2008, there was no known technique to measure the tiny amounts of $^{12}CH_2D_2$ or $^{13}CH_3D$ in a sample; it was a huge technical challenge because these two rare methane varieties differ from each other by less than a hundredth of a percent in mass. No existing instrument was sensitive enough to make the measurements, so, early in the program, we decided to build a new instrument.

Edward Young at UCLA took the lead, working with engineers at Nu Instruments in Wrexham, North Wales.[25] They adopted a conventional approach based on "mass spectrometry," which relies on separating the different kinds of methane molecules with magnets. The trick is to ionize the methane—place an electrical charge on each molecule so that it will accelerate in an electrical field, then bend the paths of those speeding molecules with a powerful magnet. More massive methane molecules travel more slowly and deflect less than do their less massive companions.

Young and colleagues pushed this mass-resolving technology to the limits. To achieve the required resolution and sensitivity, they placed a pair of large, curved metal plates and a 3-ton magnet in tandem, thus creating a fiendishly difficult problem in "ion optics." Magnets, electromagnetic lenses, and filters had to be aligned just so for the ionized $^{12}CH_2D_2$ and $^{13}CH_3D$ molecules to fly through vacuum in curving paths to hit separate targets. The resulting room-sized machine, dubbed "Panorama," was a significant risk, costing Ed Young and colleagues years of development and more than $2 million. And it worked.[26]

DCO scientists conducted the first successful experiment in which $^{12}CH_2D_2$ and $^{13}CH_3D$ molecules were simultaneously measured on November 6, 2014, at the Welsh development site. Panorama beautifully resolved the two tiny isotope peaks from a sample of commercial Welsh coal gas. The instrument was transferred to California soon thereafter, with the first studies on natural samples and the first publications produced in 2015. Some of us had been a bit nervous about the big investment, but Ed Young was confident. "I knew it would have to work," he recalls. Ed is also philosophical about Panorama: "People like to say that good science should not be instrument driven, but now and again a breakthrough in instrumentation propels us forward in our science. I think the development of this instrument is one of those instances."

Lasers

Even as DCO supported development of Panorama, we hedged our bets. Shuhei Ono, a newly minted assistant professor at MIT and a master at using traditional mass spectrometers, had been trying for several years to develop a radically new kind of isotope-measuring technique based on laser spectroscopy.[27]

Shuhei has learned to explain the principle of "quantum cascade laser spectroscopy" concisely, in just a few elegant PowerPoint slides. Gas molecules like methane absorb light at hundreds of different narrow wavelengths—a consequence of precisely tuned electron vibrations. Much as a violin's strings will resonate at certain harmonic frequencies, so, too, will an atom's electrons. Those wavelengths are extremely sensitive to the molecule's isotopic composition. Substitute ^{13}C for ^{12}C or D for H and the resonances shift dramatically.

Employing a powerful tunable laser, Ono thought he might be able to tease out the ratio of normal methane to $^{13}CH_3D$. If he could measure the intensity of absorbed light with a sensitive enough spectrometer—one that could resolve the characteristic wavelengths of the different methane variants—then he might just have an alternative technique. What's more, the laser setup is much cheaper and could eventually be miniaturized to a relatively portable version—perhaps even one that could fly to Mars someday.

The idea was great on paper, but new techniques take money, and traditional funding agencies seemed reluctant to invest in the unproven spectroscopic application. In 2012, DCO granted Shuhei a modest $100,000 grant. It wasn't nearly enough to build a new instrument, but it was enough leverage to encourage others to jump on board. Within a year, Shuhei had his instrument and the first results.

It worked better than he imagined. Sharp peaks differentiated the methane variants, while peak intensities matched well with relative quantities. Now we have two complementary techniques, and studies of isotopes in small gas molecules are exploding. New results on oxygen and carbon dioxide are already published, and the prospects for understanding Earth and its carbon cycle have never seemed brighter.

As more and more data come in, it's becoming clear that the methane story is complex and nuanced.[28] A few samples are predominantly biotic and cool; others are deep and hot. But many methane samples, including those from sources as diverse as oil wells, deep-sea microbes, and cows, display distributions of the five methane variants that suggest mixing of sources that are not completely in equilibrium. Far from being dismayed, we at the DCO now realize that the isotopic variations in methane and many other small molecules have the potential to reveal a wealth of insights on deep carbon never before imagined.

What Comes Back Up?

A lot of the carbon that goes down to Earth's mantle eventually comes back up. Some of it, in the form of enterprising carbon dioxide molecules, oozes out diffusely across large expanses of land above deep zones where heat or fluids slowly unlock the carbon in rocks. Earth's crust exhales methane too. Hidden and largely uncharted deposits of methane-rich ice, some of it buried in Arctic permafrost and even more layered in sediments of the continental shelves, release methane when they warm and melt. Microbes, termites, and cows also generate measurable amounts of methane, while all air-breathing animals produce CO_2 as by-products of metabolism. But these sources are subtle and locally modest, difficult to quantify and integrate on a global scale. By contrast, volcanoes spew out significant amounts of carbon-rich gases.

Mount Etna on the east coast of Sicily is the world's single largest point source of carbon dioxide, emitting almost 5,000 tons per day on average, with pulses approaching 20,000 tons per day during big eruptions.[29] This extreme gas production is a consequence of the thick decomposing limestone layers through which Etna's lavas pass. Many other volcanoes, even those far from carbonate rocks, generate hundreds of tons of CO_2 per day, much of it originating in the deep mantle. Taken all together, volcanoes likely represent the single largest natural source of carbon dioxide in the atmosphere. But how much?

Volcanic Carbon

All volcanoes produce carbon dioxide as they simmer and steam, some more than others. But how much CO_2 is vented to the sky? Is the rate constant, or are there carbon-rich hiccups? And what produces more carbon—the slow and steady background release or the occasional explosive event? Given the significance placed on carbon dioxide in the air, especially the role that human activity is having on atmospheric composition, hadn't we better find out whether volcanoes are a dominant source or just a blip in background levels?

From its inception, the Deep Carbon Observatory mounted a concerted global effort to document volcanic gas emissions. DCO scientists championed the development of new instruments—lightweight portable gas sensors with round-the-clock radio monitoring, laboratory tools for gas chemical and isotope analysis, even carbon-sensing drones for access to dangerous and remote locations. Gathering experts from around the world, they formed the Deep Earth Carbon Degassing initiative, or DECADE.[30] DECADE joined forces with NOVAC, the Network for Observation of Volcanic and Atmospheric Change, a collaboration of local organizations that have instrumented more than forty volcanoes on five continents.[31] Coordinating the objectives and concerns of governments on five continents is not easy, but the stakes in the volcano-monitoring game are clear and compelling.

Obtaining reliable measurements of CO_2 emitted by a volcano is exceedingly difficult. For one thing, the atmosphere already has about 400 parts per million CO_2, so volcanic contributions represent at best a modest local increase over that pervasive high background. What's more, volcanic gases are highly variable. They come in pulses from below; they swirl and oscillate with the winds. Under such challenges, direct measurements of total CO_2 from an active volcano are almost impossible.

A much more reliable estimate of total emitted carbon dioxide comes from measuring the ratio of CO_2 to a second gas, sulfur dioxide (SO_2), which is a significant (and *very* smelly) emission from many volcanoes.[32] Most of Earth's atmosphere lacks significant SO_2, so there's no background competition. What's more, SO_2 creates a strong absorption signal, so it's much easier to measure the total emission of that gas, and measurements can even be made from satellites. So, if you can determine the ratio of CO_2 to SO_2 molecules near a volcano, along with the total amount of SO_2, then it's an easy task to calculate the quantity of carbon dioxide produced by that volcano.

Outfitting an active volcano with gas-monitoring instruments can be grueling and dangerous work. Toxic blasts of hot gas, searing lava flows, boiling ponds, and the occasional flying lava bomb are routine hazards. Volcanologists don protective helmets and gas masks to deploy those instruments. They lug heavy loads of equipment and hike to the very crater edges at the summits of dangerous active volcanic cones. Scientists and their equipment are battered by wind and weather and corrosive volcanic gases. DCO's experimental station at Villarrica in Chile was lost when the volcano erupted in March of 2015. A year later the instruments at the rim of Poás in Costa Rica were destroyed as well, but not before sending data on a 100-fold increase in CO_2 emissions a few days prior to the eruption.

Volcanoes are unpredictable; they can be calm for a long time—years, decades—and then erupt with sudden fury. Volcanologists flock to these danger zones, especially at times of enhanced activity. No wonder, then, that volcanology is one of science's deadliest fields. The small community of only a few hundred researchers suffered more than twenty deaths between 1980 and 2000.

Ask a volcanologist and she'll tell you stories of the friends lost. David Johnston of the US Geological Survey died at age thirty on the morning of May 18, 1980, as his observation post 6 miles from the crater of Mount St. Helens was engulfed by the huge eruption.[33] His last radioed words: "Vancouver! Vancouver! This is it!" On that fateful day, Johnston had switched shifts with his colleague Henry Glicken, who died eleven years later in an eruption on Japan's Mount Unzen—a blast of incandescent gas and ash that also killed French volcanologists Katia and Maurice Kraft.[34] The January 14, 1993, eruption of the Galeras volcano in the Colombian Andes was even more deadly.[35] On a field trip that was supposed to be the highlight of a volcanology conference, six scientists and their companions died in a sudden explosive barrage of red-hot boulders, lava, and ash. The trip leader, forty-year-old Stanley Williams of Arizona State University, was almost killed as well. One flying boulder broke both of his legs and almost severed his right foot. A second, baseball-sized rock crushed his skull, sending bone fragments deep into his brain.

Why would volcanologists venture into these lethal environments, knowing full well that every step in the shadow of an active volcano could be their last? Without hesitation, they will explain that it's worth the risk. Volcanoes are among nature's most awesome spectacles, providing an unparalleled glimpse into Earth's deep dynamic interior. Volcanologists visit some of the remotest and most beautiful places on Earth; their scenery is breathtaking, for some a life-altering, spiritual experience. Most important, the need to understand the behavior of these restless mountains, especially to learn how to predict eruptions, is critical for the more than a half-billion people who live within literal striking distance of an active volcano. In this pursuit, carbon is the key.

Crust versus Mantle

Volcanologist Marie Edmonds has spent years studying active volcanoes—a fascination that began with a 1980 BBC television feature on the eruption of Mount St. Helens.[36] "I was only five years old," she recalls, "but I have a very vivid memory of the great swaths of trees razed to the ground by the lateral blast." Encouraged by family and teachers, she pursued both science and music (as a concert pianist).

Science won, but she wavered between Earth science and astronomy (she wanted to become an astronaut) before geology won out during her final undergraduate year.

Now on the faculty of the University of Cambridge, where she obtained both undergraduate and PhD degrees, Edmonds has led an adventuresome life. She worked as a postdoc at volcano observatories in the Caribbean and Hawaii, and participated in expeditions to many active volcanoes—the 2004–5 reawakening of Mount St. Helens, the 2006 eruption of Augustine in Alaska, and the exceptionally dangerous Soufrière Hills volcano on the Caribbean island of Montserrat. Her fieldwork was at times hazardous. The unpredictable explosive eruptions at Soufrière Hills were bad enough, but the greatest dangers came from poorly maintained helicopters. "A door fell off the helicopter, right next to me, missed the tail rotor by inches. Another time one of the engines failed. We did lots of stuff I definitely would not do now that I have children who depend on me!"

Edmonds's research focuses on the question of what comes back up. In one influential study published in *Science* in 2017, Edmonds worked with Cambridge undergraduate Emily Mason (now continuing as a PhD candidate) to study carbon emitted from the distinct family of "arc volcanoes," chains of active volcanoes that form near places where two tectonic plates collide.[37] As one plate plunges under the other in a subduction zone, wet, buried rocks heat up, partially melt, and produce the magmas that rise to form a curving line of volcanoes like Alaska's Aleutian Islands. That 1,100-mile-long chain features dozens of volcanoes, several active in any given year. All of these majestic peaks emit carbon dioxide. Edmonds and Mason wanted to know the source.

They focused on isotopes. Carbon isotopes provide part of the story. "Heavy carbon," containing a higher-than-average concentration of carbon-13, implicates a source from carbonate minerals that have broken down to CO_2 through the action of heat. Lighter carbon is more likely derived from biology—the breakdown of once living cells. But carbon isotopes alone don't provide a complete picture, because all of the heavy carbonates look the same, whether subducted carbonates that are reprocessed from the deep, or much more shallow carbonates that happen to get in the way of hot magma.

Edmonds and Mason resolved the carbonate depth problem by looking at helium isotopes: lighter helium-3 comes from Earth's mantle, whereas helium-4 is more concentrated in the crust. In many volcanic zones, including those in Italy, Indonesia, and New Guinea, they discovered a heavy-helium signature that pointed to limestone in Earth's crust, not subducted carbonates, as the major source of erupted carbon dioxide.

The implications are significant. If a large fraction of the carbon dioxide venting from volcanoes is shallow in origin, then a lot of subducted carbon is not being recycled through volcanoes. Some subduction zones may deeply bury and sequester far more carbon than had been previously thought. "We're tapping a bit of carbon that wasn't there in the original equation," Edmonds explains, "so the implication is that more carbon may be returned to the mantle than previously thought."

Predicting Eruptions with Carbon[38]

Today's Earth boasts more than 2,000 volcanoes, most of which are classified as "dormant"—an indication that they have produced lava and ash sometime in the past few thousand years but are unlikely to erupt again anytime soon. Of far greater concern are the 500 or so "active" volcanoes, which have eruptions on a regular basis: in some cases daily bursts of ash and steam; in others, fearsome explosive events every century or two. The hazards are real, but human memories are short. Why else would tens of thousands of people live in homes halfway up the flanks of Mount Vesuvius near Naples, which buried Pompeii in deadly ash 2,000 years ago? Why construct housing developments on the slope of Kilauea on the big island of Hawaii, where inexorable lava flows level fancy homes with regularity, as recently as the spring and summer of 2018?

Even scarier are explosive eruptions of incandescent gas and ash— "pyroclastic" flows that can surge down volcanic slopes at close to supersonic speeds. Major population centers from Caribbean islands to the Philippines to Seattle-Tacoma in Washington State lie directly in the paths of well-documented pyroclastic flows. Hundreds of millions of people worldwide live in the kill zones of active volcanoes. Hundreds of millions more are threatened by the more distant hazards of volcanic gases and ash, which can affect air quality and periodically disrupt air travel.

Given such impending catastrophes, it seems only prudent to keep an eye on the volcanoes that have the greatest potential for destruction. Because most volcanoes give hints that they are about to erupt, many of Earth's most dangerous volcanoes are continuously monitored by government surveys and laboratories. Typical measuring devices include seismological instruments to detect the subsurface movements of magma that precede any eruption, tilt meters that sense swellings arising from magma chambers filling near the surface, and heat sensors that record the increases in heat flow when lava rises from the depths.

Volcanoes also alter the environment in other ways, providing signals that might presage eruptions. Volcanic gases could be the key. Scientists of the DECADE project have discovered that the ratio of volcanic CO_2 to sulfur rises sharply before many eruptions. This finding underscores a recurrent theme in science: Researchers wanted to monitor carbon dioxide emissions from volcanoes to understand something fundamental about Earth's carbon cycle—the process that formed Earth's original atmosphere and that continues to shape the atmosphere today. In the course of their studies, they discovered a potentially powerful and simple approach to predicting volcanic eruptions.

Diamond Hints[39]

Diamonds, both small and rare, represent the tiniest fraction of the carbon that comes up from the depths via volcanic eruptions. Yet because they are so resilient and impervious, because they grow in Earth's hidden mantle and then carry hints of their growth environment to the surface, they are unique in the stories they have to tell about the deep carbon cycle.

Diamonds preserve two compelling clues about the cycling of Element 6: fluid inclusions and isotopes. We've already met diamonds' hoard of tiny green, red, and black mineral inclusions, but not all inclusions are crystalline. Minute, encased droplets of water- and carbon-rich liquids reveal the nature of the fluids that plunge from Earth's surface and undergo complex reactions at depths. Recently discovered inclusions in diamond match the surprising findings of experiment and theory: at mantle depths, new types of carbon-rich fluids form. And like oil and water, two very different fluids can coexist in a single inclusion. Diamonds host unambiguous evidence that some petroleum-like hydrocarbons form hundreds of miles down, beyond the realm of living cells.

The carbon isotopes that form diamonds also hold telltale clues pointing to the distant, ancient sources of their carbon atoms. The great majority of diamonds, perhaps 90 percent of analyzed stones, carry isotopic signatures typical of mantle carbon. Remarkably, a small population of relatively young diamonds (and by "young" I mean no more than a few hundred million years old) forms from "light" carbon, relatively depleted in the heavier carbon-13 isotope.[40] For any sample collected near Earth's surface, such a signal would be taken as unmistakable evidence that those carbon atoms cycled at least once through living cells. But what about isotopically light diamonds? Are they telling the same story? Did their carbon atoms once reside in cells that died and were buried, then were subducted to Earth's deep interior, there to be transformed into precious gemstones? The jury is still out, but a living source for the carbon in many diamonds would not surprise those of us who are beginning to catch a glimpse of Earth's remarkable deep carbon cycle.

Carbon in Balance

Life alters the global carbon cycle in ways that are still coming into focus. For billions of years, Earth seems to have found a balance between carbon subducted deep into the interior and carbon emitted from volcanoes—processes that help to stabilize climate and environment. But how stable is that incessant cycling? No natural law requires that the amount of carbon going down—sequestered in rocks, buried in sediments, and subducted to the mantle—must exactly equal the carbon returned to the surface by volcanoes and other less violent means. No question is more central to the Deep Carbon Observatory than this balance between what goes down and what comes back up.

—

Is Earth's carbon cycle in balance? Work by Marie Edmonds suggests that many subduction zones bury much of their carbon in the deep interior. Terry Plank, by contrast, concludes that sequestering carbon through subduction is extremely difficult—the exception rather than the rule. Which is it?

In 2015, two of DCO's most thoughtful leaders, Peter Kelemen of Columbia University and Craig Manning of UCLA, attempted to summarize all the data in one elegant diagram of the deep carbon cycle—a kind of deep Earth riff on the textbook carbon cycle figures.[41] The stylish diagram has a half-dozen red arrows, each representing an important carbon flux between the surface and depths, each accompanied by one or more small boxes with magnitudes of those carbon fluxes in megatons of carbon per year. That illustration, now employed in hundreds of DCO seminars and lectures, has become an icon of how much we still have to learn about carbon in Earth.

Not one of the arrows or their boxes is well constrained. Kelemen and Manning estimate the total carbon emitted from ridge and ocean island volcanoes to be somewhere between 8 and 42 megatons per year; the flux from arc volcanoes, between 18 and 43 megatons per year. The low estimate for the amount of subducted carbon that quickly returns to the crust and air is 14 megatons per year; the high estimate is almost five times that amount. And, most sobering of all, the calculated net carbon flux from Earth's surface to the deep interior is somewhere between a surprisingly high 52 megatons per year and a low of zero—nothing!

We see hints that Earth's carbon balance can shift. Our planet has cooled over 4 billion years, so carbonate minerals that would once have broken down under subsurface heat may now survive the subductive plunge under the cooler contemporary conditions. Life also changes the equation; it keeps learning new tricks to sequester carbon in black shales, in shell-rich limestone, in coal, and in planktonic ooze. Climates change, ocean chemistry changes, and mechanisms and rates of carbon movements change.

It's a fortunate coincidence that during most of Earth history, the total carbon going down deep by subduction has been more or less balanced by the carbon coming out of volcanoes and other sources. Consequently, life has never been at a loss when it comes to finding enough carbon for thick algal mats and dense tropical forests.

Although the jury is still out and a lot more work must be done, some scientists are now coming to the weighty conclusion that this balance may have shifted. Thanks to carbonate-forming plankton, more carbon is being buried in ocean sediments than in most previous eons. Some of that carbon may have already begun the long journey into the hidden deep mantle. And because of Earth's cooling over the past 4-plus billion years, subducted carbonate is not so easily decomposed to generate the carbon dioxide that returns via volcanoes to the surface. What goes down does not necessarily come up. The numbers are uncertain, but most calculations suggest that surface carbon may be getting buried faster and faster, at rates that could deplete the domain of life in just a few hundred million years. Don't lose sleep over this prospect; these are geologically gradual changes. But the lesson is clear. Earth's carbon cycle continues to change and to surprise.

That isn't to say we can ignore carbon-based concerns. If you're going to lose sleep over the changing carbon cycle, look not to Earth, but to ourselves.

ARIOSO, DA CAPO—
Atmospheric Change

W HEN YOU BURN ANYTHING—the Library of Alexandria, California brush, a home in Dresden, old newspapers, a Stradivarius violin—an inevitable by-product is carbon dioxide. For thousands of years, from the time when Paleolithic humans learned to control fire, our species has burned fuel to heat our dwellings, cook our food, and light our path through the nighttime darkness. For a long time, the carbon "footprint" of humans—the net addition and subtraction of carbon in the air—was neutral. We burned wood and generated carbon dioxide; new trees consumed that carbon dioxide to grow more wood.

The equation began to change with the discovery of deeply buried carbon-rich fuels. Small quantities of peat, soft coal, and oil have been exploited for millennia, though not enough to alter the atmospheric balance in any significant way. The Industrial Revolution, followed by revolutions in electricity production and mechanized transportation, were the real game changers. Surging demands for energy, coupled with discoveries of petroleum and anthracite coal, powered a frenetically evolving technological society, ushering in waves of prosperity and material comforts.

In the past 200 years, we have mined carbon-rich coal and oil by the hundreds of billions of tons.[42] Burning that coal and oil releases

about 40 billion tons of carbon dioxide into the atmosphere annually—
a quantity that dwarfs by a factor of 1,000 the outputs of all the volcanoes of the world. Humans have radically changed the equation of
what comes back up.

Truth

We should not be coy about carbon and its role in climate change.
Four facts are indisputable.[43]

FACT ONE: Carbon dioxide and methane are potent greenhouse
gases. Their molecules trap the Sun's radiation, reducing the
amount of energy radiated into space. Higher concentrations of
carbon dioxide and methane in the atmosphere mean that more
solar energy is trapped.

FACT TWO: The amounts of carbon dioxide and methane in
Earth's atmosphere are increasing rapidly. Compelling evidence
comes from varied sources, but gas bubbles trapped in polar
ice—mile-long drill cores with more than a million annual
ice layers—provide direct and irrefutable evidence of recent
changes in the air. For almost all of the past million years, carbon dioxide concentrations fluctuated between 200 and 280
parts per million (ppm), the lowest values corresponding to
episodic ice ages. In the mid-twentieth century, the value surpassed 300 ppm, perhaps for the first time in tens of millions of
years. In 2015, CO_2 topped 400 ppm. Every analysis shows an
ever-faster rise in this critical value—faster than at any time in
many millions of years.

The increase in atmospheric methane is even more dramatic. For a million years, methane concentrations oscillated
between about 400 and 700 parts per billion (ppb), again tied
to periods of advancing and retreating glaciers. In the past 200

years, that value has tripled, soaring to almost 2,000 ppb. Like CO_2, methane concentrations are higher and rising more rapidly than at any time in millions of years.

FACT THREE: Human activities, primarily the burning of billions of tons annually of carbon-rich fuels, are driving almost all of the changes in atmospheric composition.

FACT FOUR: Earth has been warming for more than a century. Record keeping begun in 1880 shows that the twelve hottest years all occurred in the past two decades. The year 2014 was hotter than any previous year, 2015 was hotter than 2014 by almost half a degree, 2016 set another new record, and 2017 was almost as hot as 2016. The average temperature in the first decades of the twenty-first century was more than 2 degrees Fahrenheit warmer than a century earlier.

Almost every scientist who has examined these compelling and unassailable facts arrives at the same unambiguous conclusion: Human activities are causing Earth to heat up. This conclusion is not a matter of opinion or speculation. It is not driven by politics or economics. It is not a ploy for researchers to obtain more funding or environmentalists to revel in hyperbolic press coverage.

Some things about Earth are true, and this is one of those things.

Consequences

The doubling of carbon in air and the consequent warming of the globe in such a short time is unprecedented. Humans are conducting a geo-engineering experiment without parallel and without a safety net; the unintended consequences have already begun to appear.

With the rise in atmospheric CO_2 has come a corresponding and predictable rise in ocean carbon dioxide and a small but devastating rise in the ocean's acidity—an alteration that weakens carbonate shells and kills coral. Some marine biologists fear a global collapse of shallow marine ecosystems.

Warming air and seas are also triggering an unprecedented loss of glacial ice, both from mid-latitude mountains at high elevations and in polar regions. Sea-level rise, already noticeable along many coasts, is inevitable—perhaps a few feet, perhaps much, much more. Such changes in ocean depths are nothing new. Over the past million years, sea level has been hundreds of feet lower during at least ten icy intervals—times when up to 5 percent of Earth's water was frozen into ice caps and glaciers. By contrast, sea levels have risen close to or above modern levels at least ten times, when less than 2 percent of Earth's water was locked in ice.

The concern today is the unprecedented rate at which glaciers are disappearing and Antarctica's great ice shelves are fragmenting. As more ice melts, the oceans become deeper; an increase of 100 feet is not without precedent. If present trends continue, hundreds of millions of people living in coastal regions could be displaced within a few centuries. Entire states (Florida and Delaware in particular) and countries (the Netherlands, Bangladesh, and some island nations of the Pacific) could all but disappear.

Warming air and oceans also affect climate. Patterns of rainfall shift, and the intensity of severe storms increases. Ocean currents that warm some parts of the globe and cool others can shift as well. A 2017 study by Daniel Scott of the University of Waterloo in Ontario modeled the variable, sometimes paradoxical effects of climate change for twenty-one former sites of the Winter Olympics, with projections to the year 2040.[44] In the twentieth century, all of those venues were consistently cold, with a greater than 90 percent chance that winter days would remain below freezing. Scott's models anticipate that nine of those snowy sites, including Vancouver in Canada, Oslo in Norway, and Innsbruck in Austria, will become unreliable, with more than a quarter of their winter days above freezing. Sochi in Russia, site of the 2014 Winter Games, fares the worst in Scott's calculations, with more than half of its wintertime days predicted to soar above 32 degrees Fahrenheit by 2040.

Ecosystems around the globe display the disruptive effects of climate change. There are, of course, benefits. Arctic regions of Greenland that for a thousand years subsisted on ice fishing during the cold, dark winter months now enjoy year-round open waters. Farmers in central Canada exploit longer growing seasons. An ice-free Northwest Passage between the Atlantic and Pacific Oceans may speed shipping across the globe. And mining companies revel in exploring for rich ores where formerly ice-covered rocky outcrops have been exposed for the first time in human history.

But other changes are troubling, benefiting no one. The rapid expansion of Saharan Africa displaces once-stable villages. Arctic communities that for uncounted centuries have remained too cold for insect pests now, for the first time in recorded history, suffer swarms of mosquitoes and blackflies in July and August. Ecozones shift northward by miles every year, potentially faster than forests and fields and migratory birds can adjust.

Scientists can anticipate, perhaps even mitigate, many of the steady incremental changes caused by a warming planet. What we cannot anticipate and what pose the greatest risk are "tipping points"—positive feedbacks that suddenly and fundamentally change the rates and consequences of change.[45] Methane, a greenhouse gas far more potent than carbon dioxide, poses perhaps the greatest risk. Almost all of Earth's methane is locked away in the crust, stored as immense layers of methane-rich ice beneath the frozen tundra and continental shelves. Exact quantitative estimates are tricky, but experts agree that the worldwide extent of methane in ice is hundreds of times greater than that of all other sources, perhaps exceeding the total carbon content of all other fossil fuels. For thousands of years that methane has been dormant, buried, a passive part of Earth's carbon cycle.

The ultimate disaster scenario—the tipping point that causes some of us to wake up at night in a cold sweat—is a global-scale positive methane feedback. Warming causes ice melting and methane release, which cause more warming and more melting. Atmospheric methane levels could soar, raising temperatures along with them. We don't know that this will happen, but if the feedback begins, it will probably be too late to stop.

Make no mistake. Whatever we do to the planet, whatever changes lie ahead, life will persist and carbon will continue cycling. But are we prepared for the coming changes?

Solutions

Humans continue to dump immense quantities of carbon dioxide into the sky—an unchecked, invisible flood that represents a global change unmatched in many millions of years of prehuman history. This is no hoax. The soaring CO_2 numbers are real, and they have consequences. Those who would deny this truth are ignorant, greedy, or both.

What is an individual to do? Leading a carbon-neutral lifestyle in our time represents a daunting challenge, for carbon emissions pervade society and thwart many of our best intentions. Would you build a giant wind turbine for "clean" energy? You'd likely strip acres of vegetation and pour tons of CO_2-emitting concrete for the foundation in the process. Would you drive an electric car? The electricity would likely come from a power plant that uses fossil fuel. Public transportation, organic farming, recycled aluminum, cloth diapers—all are worthy ways to reduce energy consumption, but all still rely in some way on carbon-based fuels. If you live in a city or on a farm, or practically anywhere in between, then you are, in all probability, a net producer of greenhouse gases.

Scientists tend to be optimists. In spite of the global change that might bring unintended catastrophe, we look for solutions and see opportunities. Peter Kelemen is such an optimist. Kelemen works at the venerable Lamont-Doherty Earth Observatory, a facility of Columbia University. Situated on the Hudson River Palisades, perched on the famed basaltic cliffs to the north and on the shore opposite Columbia's main Manhattan campus, Lamont is an oasis of research on Earth's rocks, oceans, and atmosphere.

In spite of his proximity to magnificent rock formations, Kelemen has set his sights thousands of miles away, on the majestic mountainous terrain of the sultanate of Oman on the Arabian Peninsula. There, in a sunbaked landscape that soars to 140 degrees Fahrenheit for much of the year, he studies some of Earth's strangest rocks, called ophiolites—immense chunks of Earth's mantle that should be buried tens of miles down but have somehow found themselves at the summits of 10,000-foot peaks.

Upon first impression, Peter Kelemen seems an easygoing fellow. He has a soft, gently graying beard. Old friend or new acquaintance, he smiles spontaneously, shaking your hand, speaking with a soothing, relaxed tone. He gives the sense of a casual guy with whom you'd like to take a long walk. But first impressions can be misleading.

Doing geology in Oman is not for the casual geologist. The Omani culture is welcoming to a point, but field geology can seem invasive. Having foreigners poke and prod their land raises red flags. And Kelemen wants more than just a few rocks hammered from roadside outcrops. He wants to drill deep holes and bring up thousands of feet of rock core—the ultimate geological act of penetration. So, there are understandable delays and roadblocks. Permits from the minister of lands, the minister of water, and the minister of mines must be properly submitted and approved. Local Omani drilling companies must be employed and the appropriate fees rendered. And, since no one has ever attempted to drill in these ophiolite mountains, new rules and regulations may apply; no single authority seems quite sure.

Delays mean that graduate student research is in limbo, field parties are on hold, and trips are planned and canceled. Given such administrative impediments, many scientists would have given up. But Peter Kelemen has a drive, determination, and what seems to the outside world an endless store of patience and calm. And so, after years of delays, the Oman Drilling Project has begun and is achieving historic results.[46]

Kelemen's research has reaffirmed some things we already knew. Oman's ophiolite mountains represent an unexpected block of Earth's mantle that was thrust up and over shallower basaltic ocean crust by the forces of converging tectonic plates. These mantle rocks, rich in magnesium and calcium but poor in silicon, are chemically uncomfortable when exposed to Earth's atmosphere. Specifically, they react quickly with carbon dioxide, forming elegant crisscrossing white veins of magnesium and calcium carbonate minerals.

What Kelemen and colleagues have discovered is the remarkable rate of such carbonate formation. Ophiolite literally sucks carbon dioxide out of the air, forming new carbonate minerals at an astonishing rate. You can almost watch the crystals form and grow in surface ponds and pools, as mineral-rich groundwater percolates from the outcrops. Unlike many mineral-forming processes, which are rapid only at the high temperatures of Earth's deep interior, these processes occur at room temperature (which is admittedly a lot higher on average in Oman than in your living room). These new minerals also take up more volume than the old ones, expanding the formations, perhaps explaining why the Omani mountains are still growing taller by a few millimeters per year, in spite of the fact that there is little local seismic activity.

Kelemen's mind races at the implications: carbon dioxide consumed in real time by rocks. And Oman has vast volumes of ophiolite—enough to sequester all the carbon dioxide produced by humans for hundreds of years. For the time being the Omani government wants nothing to do with such sequestration schemes. Oil, not carbon sequestration, is the basis of the country's economy. But the rocks will not go away, and the prospect for contributing to the solution of Earth's carbon crisis remains. Peter Kelemen is a patient optimist.

CODA—The Known, the Unknown, and the Unknowable

O F ALL THE DIVERSE QUESTIONS in the rich science of carbon, none is more pressing to the future of humanity than carbon's role as the element of cycles. We are able to measure atmospheric carbon to a precision and accuracy unmatched in other reservoirs. We can document the fluctuations and recent worrying rise in that parameter with equal confidence. The roles of carbon dioxide and methane as greenhouse gases, and the inevitable planetary-scale warming that must follow from their increases, are facts beyond question. Individual pleas and international treaties for moderation and for change should resonate with every citizen of the globe. A sense of urgency should pervade our lives.

And yet, in spite of the ever-expanding spreadsheet of critical carbon data, in spite of growing evidence of rapid change and its potential consequences, there is much we do not know.[47] Jesse Ausubel, the Sloan Foundation officer who shepherded the Deep Carbon Observatory's origin and rise, emphasizes the proclivity of scientists to focus on safe science. "We tend to fill conferences, magazines, and airwaves with what we know," he laments. "We much less often explore and disclose the limits to our knowledge."

It's easy to understand why researchers, whose careers depend on obtaining grants and publishing papers, tend to investigate problems and conduct experiments at the safe and secure edges of the known rather than examining the nature and extent of our ignorance. Yet articulating what we don't know—drawing those boundaries on the map of knowledge and plotting expeditions to probe those great unknowns—should quicken the pulse of any true scientist.

What are the limits to knowledge? Which questions are easy to answer, which are hard, and why? Stepping back from the issue of climate change and viewing the entire sweep of Earth's deep carbon cycle through billions of years, Ausubel catalogs three intrinsic characteristics of nature that separate the known, the unknown, and the unknowable.

Deep time, the first impediment to knowledge, is all too familiar to scientists who study planets. History fades, literally eroding away from us. Thanks to countless field expeditions that have returned samples from across the globe, coupled with the expanding analytical prowess to probe those samples down to their individual atoms and molecules, we can be confident of Earth's changeable near-surface conditions throughout most of the last tens of millions to hundreds of millions of years. Over those geologically modest timescales, most minerals do persist. They hold tiny inclusions of air and water that reveal much about the recent evolution of Earth's outer layers— atmosphere and oceans.

But hard evidence of the deeper past, 4 billion years ago and more—the time so critical to understanding Earth's formation and life's origins—is all but lost to us. The entire mineral inventory of those primordial Hadean times is restricted to a few sand-sized grains. No whiff of Earth's ancient atmosphere or drop of oceans survives. We must resort to inferences, to geochemical theory, in order to do the necessary mathematical guesswork. Even so, Earth's earliest states remain all but unknown, and perhaps inherently unknowable.

Depth represents a second hindrance to knowledge. Earth's deepest mines penetrate no more than about 2 miles; the deepest drilling, less than 10 miles. We catch glimpses of much deeper zones thanks to

volcanoes that spew out chunks of mantle rocks and minerals, including diamonds, some of which appear to come from more than 500
miles below the surface. These abyssal rock and mineral specimens hint
at the nature and extent of carbon cycling among the mantle, crust,
oceans, and air. But the vast bulk of Earth's almost 4,000-mile radius
is forever out of reach, beyond any conceivable technology to sample.

We can, of course, glean hints of the deep interior. Seismic waves
provide information on the densities and compositions of deep rocks,
as well as zones of melting and movement. Magnetic surveys reflect
the dynamics of Earth's molten outer core, while experiments on synthetic rocks and minerals reproduce a range of extreme temperature
and pressure conditions all the way to Earth's very center.

We can also imagine technologies that might push back a bit more
on the frustrating limits of what is now inherently unknowable. My
favorite futuristic tool is "neutrino absorption spectroscopy," based on
the astronomical numbers of subatomic particles that stream from the
Sun. Most solar neutrinos pass right through Earth, but theorists postulate that neutrinos of specific energies should be absorbed selectively
by different chemical elements. If we could measure neutrino energies
(which we can't—at least, not yet), then we might be able to produce a
detailed three-dimensional, CAT-scan-like image of the deep interior.

To the physical barriers of deep time and space, Ausubel adds a
third impediment to exploring the unknown: the challenge of understanding and integrating singular dramatic events in Earth history.
Evolving systems experience sudden, disruptive events that irreversibly
demarcate past from future. That devastating Moon-forming impact
by Theia when Earth was a mere 50 million years old, the origin of
life some hundreds of millions of years later, and the recent rise of
technology by humans appear to have been such singularities. Other,
less dramatic tipping points, not yet recognized, may have in the past
and may again in the future lead Earth to evolve along one of two
yet-unknown divergent paths. Such bifurcation points are inherently
difficult to predict with any certainty, yet they may represent the most
imminent danger to the human species.

—

Of the several attributes that Jesse Ausubel suggests inhibit discovery of the unknown Earth, the three just described—deep time, the deep interior, and the occasional singular disruptive event—are inherent in the physical nature of our planetary home. Other equally daunting impediments concern the sociology of science and human nature. Science is limited in what we can know because we humans are limited in the way we perceive the world.

One limitation arises because we are myopic, our minds crippled by biases and blinders. Perhaps by now you've noticed that I'm a mineralogist. I see almost any facet of Earth history—volcanoes, the deep interior, the origins of life, even the Big Bang—from that distorted mineralogical perspective.

It's true for all of us in science and in every other human endeavor as well. In our efforts to make sense of complexity and chaos, we see the world in false but reassuringly simple metaphors. Continents on shifting plates don't really "collide." Diversity of life in the Cambrian Period didn't "explode." And the evolving biosphere isn't all about "survival of the fittest." These are simplifications in time and space, reducing immensely complex physical, chemical, and biological processes to misleading phrases that conform to everyday experience.

There's an even greater challenge than dealing with our individual biases: the problem of integrating knowledge. Earth, space, life—whatever the scientific domain—demands a broadly integrated perspective. Physics, chemistry, geology, biology—all come into play in virtually every scientific topic of significance to humanity. Think about the most newsworthy concerns: deteriorating environments, dwindling mineral resources, increased spread of infectious diseases, changing climate, increasing energy demands, hazardous nuclear waste, thirsty populations, starving populations. They all require complex, interdisciplinary planning, with solutions informed by the full range of scientific evidence, filtered by myriad political, economic, ethical, and religious constraints.

Carbon science is like that—a blending of concepts and principles from every branch of research. The challenge rests in integrating many parts of the carbon story into a whole. So, perhaps when we seek to understand the limits to knowledge—the nature of the unknowable— our own human limitations need to be placed at the top of the list.

MOVEMENT III—FIRE

Carbon, the Element of Stuff

Rich in carbon, Air and Earth each play their unique roles
as they bind our lives above and below.
And yet to build our cities, drive our cars,
plow our fields, cook our food,
and make all manner of essential goods—we need energy.
And so carbon, the Element of Stuff, has other tricks up its sleeve.

Fire—energy—is the currency of industry and commerce.
Fire drives our trucks and buses, lights our streets and buildings,
heats our homes, cooks our food, assembles our machines,
and manufactures a profusion of products for a greedy, grasping world.

Carbon compounds light those fires.
Carbon, transformed by the refiner's fire,
supplies the raw essence of almost everything.

INTRODUCTION—Material World

ARTH AND AIR are not enough. Society demands a flood of diverse material goods: food and clothing, houses and factories, cars and planes, televisions and smartphones. We want stuff, and not just the essentials: tough sports gear, fine wine, comfortable chairs, resilient bumpers, soft underwear, reliable computers, tasty cupcakes, sturdy backpacks, lightweight running shoes, colorful balloons, polarized sunglasses, fluffy pillows, and firm mattresses. Finicky consumers demand novelty: Velcro, Band-Aids, Post-its, superglue, Liquid Wrench, ChapStick, Teflon, gummy bears—all products of creative carbon chemistry.

To make stuff, you need atoms in diverse three-dimensional combination: chunky masses, flexible sheets, delicate filaments, and branching arrays. You need molecules in every conceivable size and shape: chains of atoms, rings of atoms, solid blocks of atoms, and hollow cylinders of atoms. Our society craves materials with every imaginable useful property: silky, resilient, transparent, sweet smelling, absorbent, colorful, insulating, abrasive, water resistant, opaque, sticky, biodegradable, UV protective, spicy, magnetic, inflammable, dense, brittle, heat conducting and electricity conducting, sweet and salty, soft and safe.

The ever-expanding catalog of society's needs and desires creates incessant demand for equally varied atomic architectures. Each material must be meticulously engineered, carefully tailored to its specialized role at the atomic scale, for an underlying principle of chemical science is that the properties of any material depend on its atoms—its collection of elements and how those elements are bonded together.

No chemical element plays the combinatorial game of bonding to other atoms better than carbon, whose chemistry is so unfathomably rich that scientists who spend their lives studying carbon have been given their own special collective noun: "organic chemists." With an international community surpassing a million researchers, organic chemists—the experts who spend their lives playing with carbon—outnumber all other chemical scientists by a wide margin.

Electron Rules

Rules, especially those based on numbers, often seem arbitrary. Sports are a case in point. There was a time (back in the 1880s) when a field goal in American-style football was worth 5 points, while a touchdown was only 4 points.[1] A touchdown was raised to 5 points in 1897; a field goal was reduced to 4 in 1904, and then to the familiar 3 points in 1909. A touchdown was modified again to its present-day value of 6 points in 1912. Regulations for safeties, point-after attempts, and two-point conversions have undergone similar shifts over the past century. Placed in this historical perspective, the rules for football scoring seem more than a little whimsical and ephemeral, inevitably subject to additional future tweaks.

Chemistry is also a kind of game, one in which atoms are the players—an ancient dance of chemical bonding where electrons determine the score. Here's the quick version: You are an atomic winner if you wind up with exactly 2 or 10 or 18 or 36 electrons. Why those numbers? Are they, too, arbitrary? Are they subject to change in universes parallel to our own? Physicists construct elaborate explanations of these "magic" numbers. It's just the rules of the game, but in this case, the rules are built into the very fabric of our Cosmos.

Some atoms are born lucky, with exactly 2 (the element helium), 10 (neon), 18 (argon), or 36 (krypton) electrons. These special atoms spend their lives as isolated, free-floating loners—solitary "inert" gases, because they don't need to rely on any other atoms to achieve their winning numbers of electrons. Other atoms just miss a magic number: Sodium, element 11, starts off with 11 positive protons and 11 negative electrons, but it readily gives up an electron and becomes a positively charged sodium ion. Chlorine, with 17 protons and electrons, just as readily takes that unwanted electron from sodium to become a negatively charged chlorine ion. Positive sodium ions attract negative chlorine ions, mingling to create exquisite, tiny, cube-shaped crystals of table salt—sodium chloride.

The periodic table features plenty of nonmetallic elements like chlorine or oxygen (with 8 electrons, 2 shy of the magic 10), more than willing to grab an extra electron or two from overstocked metals like sodium or magnesium (with 12 electrons). Most of the elements in the periodic table adopt this kind of strategy, either giving away electrons or snapping them up to win the bonding game. That's a good thing. If all atoms were equally happy with their original allotment of electrons, then there would be no reason to shuffle and share those electrons, no way to form chemical bonds, and no pathway to our exuberantly varied material existence.

In this world of mutually beneficial electron bartering and usually friendly takeovers, carbon holds a unique place as Element 6, smack in the middle of the periodic table, halfway between magical 2 and 10. Like a weary swimmer on a lake, treading water equidistant between two shores, carbon doesn't "know" just what to do. Should carbon go one way, seeking 4 more electrons to reach the magic number 10? Or should it head in the exact opposite direction, giving up 4 electrons to wind up at the magic number 2?

This ambiguity gives carbon a bonding advantage unknown to most other elements. Unlike one-trick sodium, which invariably gives up a single electron, or chlorine, which just as readily seizes that one extra electron in its struggle for atomic contentment, Element 6 enjoys many contrasting chemical roles—adding, subtracting, or sharing electrons in combinations that lead to vastly more varied chemical compounds than all of the other 100-plus elements combined. That's why carbon can create both the hardest of all known materials and the softest, the most vivid and varied colors and the blackest possible black, the most slippery lubricants and the stickiest glues.

Flammable Carbon[2]

We need stuff, but making it demands a lot of energy. As often as not, that energy comes from the heat of a carbon-stoked fire. We're in luck, for Earth holds abundant reserves of small, flammable carbon-bearing molecules—hydrocarbon-rich fuels, including coal, oil, and natural gas. Hydrocarbons, the simplest products of organic chemistry, are ubiquitous molecules on Earth and in space, each one crafted from a sturdy backbone of carbon atoms and decorated with a halo of hydrogen atoms. Natural gas, or methane, is the simplest hydrocarbon—one carbon atom surrounded by a pyramid of four hydrogen atoms. Carbon contributes 6 electrons, each of the hydrogen atoms adding one more for a "magic" total of 10 electrons.

Novelty emerges when two carbon atoms link to each other, embraced by up to six hydrogen atoms around their periphery—the fuel called ethane. In that simple, flammable molecule, each carbon atom shares electrons with four neighbors, so each carbon atom enjoys 10 electrons at the same time that each hydrogen atom sees its desired complement of 2 electrons. In ethane, every atom is content.

Let's keep building: three carbon atoms in a row provide the backbone for propane, the common rural fuel stored in big, white tanks. In propane, eight hydrogen atoms surround a little row of three carbon atoms.

With four carbon atoms, a new wrinkle emerges: those four atoms can arrange themselves in two different ways, as two distinct "isomers." Butane's four carbon atoms can line up in a neat little row (the fuel butane, as employed in most disposable cigarette lighters), or they can form a T-shaped molecule, isobutane, which finds everyday use as a safe refrigerant. Pentane with five carbon atoms and twelve hydrogen atoms can form a five-carbon chain, a four-carbon chain with one side branch, or a symmetrical cross. Octane, an eight-carbon ingredient in gasoline, has a remarkable eighteen isomers with distinctive topologies (one isomer, the basis of gasoline's "octane rating," has a chain of five carbon atoms with three little side branches). Candle wax expands on this theme, boasting a hydrocarbon chain of twenty to forty carbon atoms—the more carbon atoms in the chain, the higher the melting point of the wax.

With five or more carbon atoms, new opportunities arise: carbon atoms can circle around into an array of elegant ringlike molecules.[3] Benzene, once used as an industrial cleaner but now known to be a dangerous carcinogen, features a six-carbon ring with six hydrogen atoms radiating like spokes of a wheel. Sometimes carbon rings are nested together—a motif found in most black soot particles. Naphthalene, the most common "polycyclic" hydrocarbon, pairs two of these rings. Anthracene, a common ingredient in the sooty smoke from charcoal fires and diesel fumes, has a neat line of three linked hexagonal rings, while pyrene, another soot ingredient, features a tight cluster of four rings.

Sometimes molecules adopt elaborate architectures, with fancy combinations of chains, branches, and rings—a theme displayed by many of life's vital molecules, from steroids and vitamins to the building blocks of the genetic molecules DNA and RNA. These carbon-based molecules can grow larger and larger, with dozens of five- and six-membered rings, themselves linked in chains, rings, and clusters. Indeed, there is literally no end to the variety of hydrocarbon molecules—and furthermore, most of them burn.

Coal, oil, and natural gas all form primarily from hydrocarbons. These "fossil fuels" have transformed society and, for good or ill, remain our cheapest and most abundant sources of chemical energy.

Not so on many worlds beyond Earth, where hydrocarbon molecules would make a lousy fuel. The surface of Saturn's giant, frigid moon Titan is subjected to hydrocarbon rains that pummel the surface in great storms.[4] Titan's rivers flow with methane and ethane; immense lakes are filled with the stuff. You could sail a boat on Titan's methane lakes, yet if you could light a match on Titan (which you can't, because there's no oxygen to trigger the flame), nothing would happen. Lacking a chemical oxidant in Titan's atmosphere, the hydrocarbon rain would simply extinguish any flame.

On Earth, the essential difference is abundant atmospheric oxygen, the dangerously reactive by-product of photosynthesis. Oxygen is an element exceptionally greedy for electrons. Unlike the case for any other planet or moon in our solar system, lighting a match here on Earth and in the vicinity of volatile hydrocarbons like natural gas has dramatic and dangerous consequences. In a fiery, explosive oxidation-reduction chemical reaction, hydrocarbon molecules rush to give away electrons, reacting with oxygen to produce two familiar simple chemicals: carbon dioxide and water. These rapidly flaming chemical reactions produce copious amounts of heat and light—the paradoxical "good servants" and "bad masters" of fire. For thousands of years, humans have lived with the essential benefits and ever-present dangers of open flames.

Heat is the key to making new materials, and fossil fuels are one vital source of that refining heat. Most hydrocarbon fuels burn at high temperatures, close to 3,600 degrees Fahrenheit for natural gas on your stove or butane in your cigarette lighter (your Thanksgiving turkey, by contrast, bakes at a mere 400 degrees Fahrenheit). Some specialized tasks, such as metal welding and cutting, require the much hotter 6,000-degree flame produced by an oxyacetylene torch. Nevertheless, when it comes to unearthing carbon-rich fossil fuels to create the diverse world of stuff, burning is not always their most practical use. Coal, petroleum, tar sands, and natural gas often find much better employment as the starting points for making most of the materials in our lives.

Diverse Carbon

Hydrocarbons are a tiny domain in the expansive kingdom of organic chemistry. Earth manufactures millions of different kinds of carbon-rich molecules, but what drives this profusion? The answer lies in carbon's unique ability to bond to many different elements—dozens of diverse chemical elements in the periodic table, including itself. Most carbon compounds in your body incorporate oxygen, while nitrogen, sulfur, and phosphorus are also typical of life's most critical building blocks. Carbon readily bonds to metallic elements like iron, titanium, and tungsten in tough "carbide" machine parts and abrasives, as well as to nonmetals like fluorine and chlorine.

The resulting carbon-based chemicals provide the foundation for modern industry—an industry that relies on hydrocarbons as the essential raw materials. And here we come to an important realization: Eventually, we must stop burning coal and oil. Cite the environment if you will—and to be sure, concerns about effects of burning fossil fuels on the environment are real—but the simple truth is that carbon-carbon bonds in coal and oil will eventually become much too valuable to burn. These ubiquitous bonds between carbon atoms are the most essential chemical characteristic of our material world; they lie at the heart of almost every product that we consume in our lives. Our planet has provided us with many alternatives for energy: abundant sunlight, incessant winds, renewable biofuels, pounding waves, inexhaustible geothermal heat, and the energetic nuclear reactions of radioactive uranium. Carbon-based "fuels," by contrast, represent the indispensable feedstocks of our blossoming material world.

SCHERZO—Useful Stuff

Hot

Fire! Fire is the key to refining coal and oil, themselves rich remains of buried life baked and compressed into complex mixtures of countless thousands of different molecules.[5] Most molecular ingredients are small, familiar hydrocarbon fuels: methane, ethane, propane, octane. Many others are large, complex molecules boasting dozens of carbon atoms with crucial bonds to oxygen, nitrogen, sulfur, and other elements.

The trick to refining is heating this smelly, black chemical stew in a tall, segmented cylinder, much hotter at the bottom than at the top. You see these distinctive "distillation columns" every time you drive by chemical facilities with their bristling ranks of towering metal tubes, some flaming at their tops as they burn off small amounts of excess methane—a trivial quantity of natural gas that chemical engineers deem too meager to collect for profit. Each of those towers performs multiple steps in the chemical separation process.

As the mixture of ingredients is heated in the column, each molecular component reaches its boiling point at a different height. A complex system of pipes juts off of the column at successive levels—each pipe drawing away and collecting a different product, thus distilling the complex, black liquid in stages. Smaller molecules generally boil at lower temperatures, and thus they emerge from stages higher up in the column—propane and butane near the cooler top, gasoline and kerosene in the middle, with thick, sticky liquid asphalt and waxes flowing out of the hotter bottom. Refineries link their distillation columns in a carefully choreographed chemical dance, with each column accomplishing some phase of the essential task of selecting and concentrating critical organic chemicals.

Once those diverse carbon-based molecules are distilled and purified, chemical tricks to produce new compounds abound: mix varied chemicals in large vats and retorts, stir them, squeeze them, add just the right reactive ingredients, perhaps add a pinch of catalyst, then cook at just the right temperature. Diverse recipes yield cookbooks full of useful synthetic products. Among the most important materials for modern life are polymers—plastics with nicknames like PETE and PVC, the synthetic fibers nylon and rayon, paints, glues, rubber, and hundreds of other chemicals that play innumerable roles in our lives.

All of these materials feature countless small molecules linking end to end to form long chains with carbon backbones. Life, too, has learned these chemical tricks: skin, hair, muscle, tendons, and ligaments—all are biopolymers. The same is true with leaves and stems, roots and wood, filaments of algae and strands of spider silk. Clever chemical manipulations also yield myriad carbon-based compounds, such as waxes and resins, fats and oils, lubricants and glues, cosmetics and drugs.

Check out the nutrition label on your favorite snack food. Everything we eat is rich in carbon-based molecules: amino acids, the building blocks of protein; lipids, the components of fats and oils; carbohydrates, including sugar, starch, and dietary fiber. Carbon provides the carbonated fizz of soda, and the alcoholic buzz of booze.

Let's explore some of the properties of carbon compounds that make them so indispensable to everyday life.

Cold

Carbon chemistry, which brings us the hottest flames, also provides the most efficient portable source of refrigeration. Carbon dioxide, CO_2, freezes at 109 degrees below zero Fahrenheit into a colorless solid commonly called "dry ice." The ice is called "dry" because of its unusual property of subliming directly from solid to gas. Unlike water, carbon dioxide has no liquid phase, at least not at room pressure. So you can't pour cold carbon dioxide from one container to another, but you can carry it around in convenient frigid chunks.

Dry ice finds several quotidian uses, most commonly in the food industry. Without the help of a mechanical refrigerator, you can enlist CO_2 to freeze food, carbonate beverages, make ice cream, and ship perishable goods. Frozen carbon dioxide also finds clever uses in bug control because mosquitoes and bed bugs are attracted to CO_2; they cluster around the dry ice and freeze to death. Plumbers use portable packings of dry ice to wrap around copper pipes, thereby creating plugs of frozen water when no shutoff valve is available. Doctors freeze and remove warts; environmental engineers freeze and clean up oil spills; firefighters use dry-ice pellets and carbon dioxide extinguishers to fight fires, at once cooling and suffocating the flames.

Frozen carbon dioxide also finds a unique role in stagecraft to produce fog. Dump pieces of dry ice in water and you'll generate a dense, ground-hugging fog perfect for eerie nighttime effects. What the audience doesn't see is how cold and wet that fog can be, as the cold, subliming CO_2 lowers the dew point and saturates the denser cold air with water vapor. I recall playing in a small band for a show once when an overzealous dry-ice effect produced copious fog that spilled over the lip of the stage and quickly filled the orchestra pit. For a time we were literally playing blind. The clammy, water-saturated air condensed on everything, leaving a slippery wet coating on floor, chairs, music stands, and instruments.

Sticky

The properties of materials depend on their atoms and how they bond together. Take glue, for example. The hallmark of a good adhesive is that it sticks to almost everything. At the scale of atoms—the scale where most material properties originate—"sticking" refers to the strong attraction of positive and negative electrical charges. The molecules of glue must have unusually strong surface charges, often realized by the strong negative electrostatic charge of OH^- "hydroxyl" groups that are bonded to carbon atoms. When a carbon-based molecule has multiple hydroxyl groups projecting outward, they can induce a corresponding positive charge at almost any surface. Positive attracts negative, and voilà! the molecules stick.

Nature is replete with stickiness, through adhesives that invariably incorporate a carbon backbone.[6] Geckos climb walls with hydroxyl-decorated feet. Venus flytraps excrete sticky, hydroxyl-covered mucus that ensnares insects. Mussels and barnacles attach to the hulls of ships in much the same way, costing shipping companies a small fortune in annual cleanings, not to mention the associated lost time of valuable vessels staying idle in port. Every year sees the introduction of new nonstick marine paints, with molecular configurations that limit, but can never completely eliminate, these electrostatic tricks.

Stickiness is big business. The adhesive industry feeds a multibillion-dollar annual market, with customers as diverse as aircraft and automobile manufacturers, construction companies, online retailers, and medical professionals. High-tech glues and sealants now replace metal welding, thereby speeding construction and reducing weight. They hold in place the glass of skyscrapers and the windshield of your car. Everyone relies on adhesives, in products like disposable diapers, false teeth, Band-Aids, hearing aids, postage stamps, envelopes, and Post-its, as well as for dozens of everyday tasks, from wrapping birthday presents to gluing broken furniture.

Superglue epitomizes the quirky nature of discovery in the world of adhesives.[7] Organic chemist Harry Coover Jr. and a team of researchers at the Goodrich Company stumbled upon the first of the family of superglues, a small, carbon-based molecule called "cyanoacrylate," in 1942 as part of the war effort. They were trying to develop an improved clear-plastic gunsight; cyanoacrylate, which stuck to everything it touched, was quickly rejected.

Fast-forward to 1951. Coover, having moved to Eastman Kodak and working with chemist Fred Joyner, realized that supersticky cyanoacrylate might prove to be a valuable adhesive. Kodak agreed, and the first superglue hit the market in 1958 as "Eastman #910." Numerous variants and competing brands soon emerged; all, however, relied on the property of cyanoacrylate molecules to remain in liquid form when stored in a sealed container but to bond strongly together when exposed to water or atmospheric humidity.

Nontoxic superglues, with their ability to bind to many different kinds of surfaces and solidify in varied environments, have found dozens of novel uses beyond the familiar repair of broken objects or assembly of parts. Marine biologists and aquarium enthusiasts use superglue to attach living coral fragments to rock. Superglue vapors stick to oily residues on smooth surfaces to produce delicate fingerprints for forensic analysis. And superglue has become the go-to adhesive for applications to skin and bone. It finds frequent use in the strengthening and repair of calluses for athletes, rock climbers, and string players. And doctors and veterinarians use superglue for mending bones and closing wounds, especially in emergency situations when gluing is faster and safer than traditional suturing or stapling.

Slippery

Slippery molecules, in contrast to glues, minimize their surface charges. Lacking an electrostatic attraction, the molecules simply flow over each other like uncooked grains of rice. Waxes, grease, and oils create slippery surfaces because they are formed from hydrocarbon molecules—carbon atoms surrounded by hydrogen atoms. Every atom in a hydrocarbon molecule has fully satisfied its magic number of electrons; none of the atoms in oil or grease are looking for other atoms to bond with.

Everyone has stories of unfortunate slips—irreversible instants when something bad happened. I think back to one such vivid moment from my former musical life. Sanders Theatre, Cambridge, Massachusetts, February 1975—the third or fourth rehearsal of a new chamber work, *Eh, Joe*, by Harvard composition professor Earl Kim.[8] Fiendishly difficult, the piece had three brass parts and three string parts (try balancing those instruments!), plus a speaker who had to master a singsong narrative of poetry by Samuel Beckett. Twelve hours of rehearsals for a twenty-minute piece. We sat in a semicircle, trumpeter Ken Pullig and trombonist Stan Schultz to my left, wunderkind undergraduate cellist Yo-Yo Ma to my immediate right, conductor Kim and Sprechsinger Lois Smith in front. The piece proved challenging, finicky. With an upcoming performance and recording, rehearsals were long and intense.

An hour into an onstage rehearsal, while we were tackling a tricky ensemble passage for perhaps the fifth or sixth time, a deafening crash to my right. We all stopped playing. What had happened?

Then we saw the 16-inch metal wrench that had slipped from the greasy hand of a lighting tech. He had been working high above us on a catwalk, adjusting a canister light, forced into an awkward position, when gravity and grease betrayed him.

The massive wrench had landed 6 inches to the right of Yo-Yo. We were all in a state of shock, concentration broken at the near-fatal miss. Literally shaking, we could not go on. The rest of the rehearsal was canceled as we were reminded of the irreversibility of the occasional very bad moment—the inevitability and unpredictability of death. We left the stage, the overtones of Beckett's fatalistic poetry ringing in our minds.

Yo-Yo Ma was visibly stunned. In a quaking voice he spoke softly: "It almost hit my cello!"

TRIO—Nano Stuff

Novel

Change, transformation, and endless variety are hallmarks of carbon chemistry. Consider graphite, with its tiny triangular planes of atoms that form superstrong flat atomic sheets. Those individual atomic layers are almost indestructible, not unlike a thin, tough sheet of plastic, but the forces between adjacent sheets are weak and readily broken. As a consequence, graphite layers slide across each other as easily as the papers stacked on your desk are scattered by a gentle breeze.

If graphite is among the softest known materials, its individual carbon sheets are among the strongest and most resilient. But of what use is a single sheet of carbon atoms? How could you capture and study such a nanoscale material? For decades, long before they had a sample to study in the lab, scientists had speculated about the unique electronic and mechanical properties of this enigmatic substance, dubbed "graphene." Perhaps graphene would be a semiconductor, or possess unusual magnetic properties, or provide a superstrong material for nanoengineering? But it was not until 2004 that University of Manchester researchers Andre Geim and Konstantin Novoselov made the graphene breakthrough. Their Nobel Prize–winning, high-tech solution to isolating single carbon layers? Scotch tape![9]

Now, anyone can isolate and investigate graphene layers. Start with a nice flat crystal of graphite. Apply a piece of Scotch tape to the flat surface, and then remove. Typically, a few graphite layers will come away, but repeated application of the adhesive tape does the trick. Eventually, all but one atomic sheet remains. Dissolve away the adhesive and you are left with a perfectly flat single atomic layer of carbon.

The ensuing flood of discoveries—more than 10,000 scientific publications per year and still growing rapidly—holds promise for a host of revolutionary technologies.[10] Graphene layers are transparent and strong, so they could play roles as microscopic windows in nanoscale devices, as composites for the engineering of artificial skin and bone, and as materials for a new generation of ultrathin condoms. The layers are insoluble in water, so graphene coatings can protect the surfaces of otherwise soluble or easily corroded devices. Remarkably, water can still "wet" the surface of a coated object, since H_2O molecules can still interact through the holes in the single layer of carbon atoms. Thus, graphene could protect a wide range of water quality, humidity, and biological sensors.

The possibilities for electronic applications are no less intriguing. Semiconductors control electrons, changing their flow rates while switching them from one path to another. Our modern electronic age is utterly dependent on devices made primarily of silicon semiconductors—diodes, transistors, and integrated circuits. Graphene is now challenging silicon's supremacy. The first graphene transistor—the essential workhorse of the electronic age—was demonstrated in 2004, with rapid subsequent developments. In 2008, a team of German researchers constructed a graphene transistor only ten atoms wide—by far the smallest so far, and close to the theoretical limit. A variety of integrated circuits soon followed, as did tiny transistors with switching speeds faster than that of silicon. What's more, such devices can be fabricated with 3-D printing, they are flexible, and they can operate underwater. Some advocates predict that graphene may soon replace traditional silicon semiconductors in many applications.

New ideas continue to flood in. Graphene has superlative thermal conductivity and could play a role in varied applications where electrical circuits must be cooled. Graphene's transparency and conductivity could be ideal for flexible touchscreens and displays. Other applications—to fuel cells, batteries, high-tech lenses, pressure sensors, and water filtration—are now in development. New variants of graphene with stacks of two or three carbon sheets, or sandwiches interleaved with other layered materials, possess unique properties that provide yet more opportunities for discovery. And one research group is even developing graphene-based hair dyes that can keep your head cooler in summertime heat.[11]

Hollow

Perhaps the most obvious attribute of graphene is its amazing tensile strength. The strength of materials comes in three flavors: "Compressive strength" measures resistance to crushing, "shear strength" measures resistance to twisting, and "tensile strength" measures resistance to pulling apart. Some materials, such as a stack of bricks or a pile of lumber, are strong under compression but weak when twisted or pulled. Other materials, like a steel chain or nylon rope, are strong under tension, but have virtually no strength under compression or shear. A few everyday composite materials, like reinforced concrete, fiberglass, and plywood, combine characteristics of two or more materials to enhance all three types of strength.

Flat sheets of graphene can't withstand twisting and easily fold when squeezed, but under tension, graphene is unmatched; it is a hundred times stronger than the strongest steel wire and has twice the tensile strength of diamond. The reason for these extreme characteristics lies in the nature of carbon-carbon bonds. Diamond, with each carbon atom sharing its electrons among four neighbors, is the strongest known three-dimensional crystal. Carbon atoms pack so tightly that diamond has the highest electron density (electrons being the key to chemical bonds) of any material at Earth's surface. The distance between adjacent carbon atoms is only about 6 billionths of an inch— much smaller than in most other crystals. That's why diamond is so tough and hard. But in graphene layers, the carbon-to-carbon distances are even shorter—just 5.5 billionths of an inch. That's because each carbon atom shares its allotment of four bonding electrons with only three neighbors. The electrons within the layers are even more tightly packed, and the resulting bonds are even stronger and shorter than in diamond.

Graphene's superior tensile strength suggests a clever strategy for carbon nanoengineering. Layers make lousy ropes and wires, but what if you could curl a carbon sheet around on itself and make a tiny, hollow cylinder? Then you'd have an incredibly strong carbon "nanotube."[12] Many variants are possible: different tube diameters, single tubes, double tubes, and nested tubes with multiple concentric cylinders.

Carbon fibers of various sorts have been known and studied since at least the 1950s, but research exploded after a 1991 discovery by Japanese physicist Sumio Iijima, who produced abundant carbon nanotubes by passing strong electrical currents through graphite. Given this new reliable supply, more than 100,000 scientific articles and 10,000 patent applications have been submitted.

The strength of hollow carbon nanotubes is astonishing: a thread just a twentieth of an inch in diameter can support more than 10 tons of dead weight. The engineering potential for designing lightweight bridges, buildings, aircraft, and a new generation of composite materials is mind-boggling. Science fiction writers have jumped on the concept, positing space elevators that shuttle people and supplies to fixed orbiting platforms hundreds of miles above Earth's surface on cables of carbon nanotubes. Even with such futuristic potential, the allure of nanotubes goes beyond their strength. A flood of applications in manufacturing, energy supply, electronics, and medicine continue to occupy an army of scientists around the world.

Clever[13]

Graphene layers and nanotubes point to other possible forms of carbon. Seal off both ends of a nanotube and you can make a delightfully varied array of enclosed forms, including the soccer ball–like, sixty-atom "buckyball," or more elongate football-like molecules with seventy or more carbon atoms. These elegant forms of carbon are collectively known as "fullerenes," in tribute to the geometrically similar geodesic domes of American architect Buckminster "Bucky" Fuller.

The existence of fullerenes was predicted more than a half-century ago, though it was not until 1985 that a team of scientists from the University of Sussex in England and Rice University in Texas described a reproducible path to their synthesis and analysis.[14] The Nobel Prize–winning discovery was followed by discoveries of fullerenes in candle soot, smoke from forest fires, lightning discharges, and even cosmic dust surrounding distant, carbon-rich stars. Intense research on these cage-like molecules has led to numerous new forms: nano-onions featuring nested cages within cages, dumbbell shapes with two buckyballs linked by a carbon chain, and carbon containers enclosing a host of smaller atoms or molecules.

Given the basic forms of flat graphene layers, hollow nanotubes, and enclosed fullerenes, it's only a small step to imagining more exotic geometries. Nanobuds look like little rounded bumps on a nanotube or on a larger fullerene. Nanotubes can connect to each other at right angles in nanojunctions, or they can project vertically from a graphene layer in nanopillars. Buckyballs can fill nanotubes like peas in a peapod, while concentrically nested nanotubes can elongate or contract like the shaft of a compact umbrella. You can even postulate nanotubes curving around a circle to make a perfect doughnut-shaped molecular torus.

Armed with such an array of shapes, scientists and inventors dream of a new generation of molecular machines—nanoscale levers, pulleys, wheels, and axles.[15] Thanks to carbon nanotechnology, the atomic-scale motors, electrical circuits, and electronic components that are needed for the next generation of implanted medical devices, drug-delivery micro-containers, and molecular-scale computers seem almost within reach.

SCHERZO, DA CAPO—Stories

"*PLASTICS!*"—THAT IS the "one word" whispered to Dustin Hoffman's aimless character, Ben, by Mr. McGuire in Mike Nichols's 1967 film *The Graduate.*[16]

"Exactly how do you mean?" Ben replies.

"There's a great future in plastics. Think about it. Will you think about it?"

That memorably amusing and clueless scene contains more than a kernel of truth. Plastics, or "polymers," have changed the world. Polymerization is a chemical reaction in which numerous small molecules, or "monomers," link into a chain or network to form a "macromolecule," a single extended molecule with thousands of atoms, almost always with a backbone of carbon atoms. Natural polymers are part of every living thing: wood, hair, muscles, spider silk, skin, leaves, tendons—the list goes on and on. Given the ubiquity of polymers in biology, chemists were rather slow to emulate, and ultimately improve upon, nature.

Rubber, first employed by Mesoamerican cultures in its natural form more than two thousand years ago, became one of the first polymers to attract the scrutiny of the chemical community.[17] Natural rubber comes from the rubber tree's curious latex sap, a remarkable product that hardens into a flexible waterproof material that can be molded into sheets, balls, and other useful objects. Yet, in its uncured state, as harvested directly from nature, it possesses a litany of undesirable qualities: natural rubber is too sticky and too smelly, becomes runny when too hot, and is brittle and cracks when too cold. The reasons for its laundry list of properties, both desirable and undesirable, lie in the structure of the rubber polymer. The long, strong carbon chains of rubber molecules can slide past each other, providing both strength and flexibility, but under only a narrow range of temperatures.

The modern industry devoted to polymers, including the vast and growing class of materials called plastics, began with the invention of vulcanization in the 1830s—an innovation claimed by rival American and British chemists. Vulcanization is a chemical process by which sulfur or some other chemical added to a polymer forges strong cross-linking bonds—a kind of molecular cross-bracing. The result is a much harder and more durable material (and less smelly, to boot). In the case of rubber, the process of adding sulfur to the sticky sap of the rubber tree and curing the mixture with heat led to the vastly improved products we use today: gloves, galoshes, pencil erasers, hoses, rubber bands, party balloons, inflatable boats, and, of course, tires for every sort of wheeled vehicle. A few more additives yield much harder rubber varieties employed in football helmets, skateboard wheels, bowling balls, and clarinets.

The turbulent years following World War I saw a remarkable transformation in chemistry, as chemists began increasingly to think of materials at the scale of atoms, with an eye to molecular architecture. In 1920, German chemist Hermann Staudinger revealed the nature of polymers as giant molecular chains with carbon atoms forming a sturdy backbone—work for which he would receive the Nobel Prize in Chemistry decades later.[18] Staudinger's recognition of varied natural biopolymers, including rubber, proteins, starch, and cellulose, proved the ubiquity of macromolecular substances. He also predicted that synthetic polymers would someday be developed with properties rivaling those of natural materials.

In spite of Staudinger's discoveries and his accurate vision of the future, synthetic chemists were at first stymied. In the mid-1920s, researchers were still hard pressed to create macromolecules more than a few dozen monomers in length—too short for any useful applications. To be sure, a few novelties had emerged. The Belgian-American chemist Leo Baekeland had experimented with heated mixtures of the common chemicals phenol and formaldehyde to produce synthetic shellac (a substance previously obtained almost exclusively from the excretions of Asian lac beetles).[19]

In 1907, Baekeland refined his synthesis methods, producing the first plastic, dubbed "Bakelite"—a product employed in modest amounts for eclectic uses, including colorful and highly collectable kitchenware, toys, and jewelry. Five years later, Swiss chemist Jacques Brandenberger introduced cellophane, a flexible, waterproof film formed by reconstitution of the cellulose of trees and other plants. Commercial success followed when the Whitman candy company selected cellophane to wrap individual chocolates in their famed Whitman Samplers. Nevertheless, fundamental advances in polymer chemistry research lagged until an explosion of discoveries in the 1930s yielded new materials that have changed the world.

Polymerization (Strong)

Silky

It was in a climate of chemical optimism and discovery that the brilliant young chemist Wallace Carothers made his mark.[20] With advanced degrees from the University of Illinois and a year of teaching at Harvard under his belt, Carothers was lured to the research laboratories of chemical giant DuPont in Wilmington, Delaware. Convinced that commercial breakthroughs would come from fundamental research, DuPont hired Carothers in 1928 to lead a group tasked with pursuing "pioneering research" in polymer chemistry. Carothers and colleagues made rapid progress, creating the first synthetic rubber "neoprene," the familiar material of elastic knee braces and flexible diving suits, in 1930.

Carothers's most significant breakthrough was the February 1935 invention of nylon, a remarkable polymer that can be heated, melted, and formed into fibers, films, or a wide variety of shapes. It was first introduced as a novelty, forming the bristles of souvenir toothbrushes distributed in 1938, and then in women's stockings at the 1939 New York World's Fair. However, nylon saw increasingly widespread military applications during World War II, notably as a replacement for silk in parachutes. As the number of applications exploded, nylon made DuPont hundreds of billions of dollars in profits.

Wallace Carothers did not live to see this success. Plagued by depression, suffering from the death of his sister, feeling his personal life a failure and his chemical inspiration waning, he took his own life by swallowing potassium cyanide two days after his forty-first birthday.

Carothers was a meticulous chemist. For years, he had kept a capsule of cyanide attached to his watch chain. He knew the effects of cyanide poisoning well: the C≡N cyano group blocks your cell's use of oxygen. Such a simple molecule: two atoms—one carbon and one nitrogen—both essential for life. Yet a carbon atom linked by a triple bond to a nitrogen atom causes death, as the heart and central nervous system shut down. Ever the creative chemist, Carothers mixed his cyanide capsule with lemon juice—an acid that accelerates the poison's effects.

Foamy

At George Mason University, where I teach a course on scientific literacy for undergraduates, my demonstration of polymerization is simple, based on a safe and cheap chemical kit available through any of a number of educational supply companies.[21] Condensation polymerization is a common industrial chemical reaction whereby numerous monomers, each a small, carbon-based molecule, bond end to end to form a long, chain-like polymer. In the case of condensation polymerization, the formation of each new chemical bond releases a small molecule, usually water or carbon dioxide.

Nothing would have gone wrong if I had followed the instructions. First, pour two liquids—one clear, the other amber—into a plastic cup. Then stir the liquids until they are well mixed. Wait two or three minutes. The reaction begins slowly as a gentle yellow froth appears, but it accelerates as foam billows over the top and sides of the cup, flooding the table, sticking to anything it touches (fingers included). The foam arises from carbon dioxide molecules that are released by the condensation reaction. The whole gooey mess gradually hardens into a rounded, durable mass of polyurethane, a material ideally suited for the snug packaging of delicate electronics and the efficient installation of building insulation in tough-to-reach cracks and hollows.

It was, in retrospect, an extremely bad idea to mix the liquids in a plastic water bottle and screw on the top—especially when I hadn't tried it before. The experiment started off well enough, but when the usual foaming slowed and stopped, it dawned on me that a lot of pressure must be building up in the thin-plastic bottle. I had, in effect, assembled a small explosive device in front of my class, and the pressure was building fast.

This was dumb. Don't try it at home.

What to do? Relieving the pressure as quickly as possible seemed logical, so I started to loosen the top and . . . BLAM! The screw cap blasted straight up, bouncing off the ceiling and rebounding a couple of feet to my left. That missile was followed by a remarkable projectile blast of polyurethane, suddenly liberated by the pressure release. A yellow mass shot up 25 vertical feet and splattered the ceiling tiles with gobs of sticky, yellow insulation. (Remnants of the mishap were still there the last time I checked the Enterprise 80 Lecture Hall.) Fortunately, no one was hurt, though a few students in the first couple of rows were unexpectedly adorned with little yellow dabs of sticky foam.

Mistaken

Proteins are ubiquitous biopolymers composed of a linear chain of amino acids, each a small, carbon-based molecule. The structure of a protein is determined primarily by the exact sequence of the twenty varied amino acids employed by biological systems. Connected end to end in a sequence, amino acids can form flat sheets (as in cartilage) or strong fibers (hair and tendons) or more random coiling shapes. Those twenty amino acids, strung together by the hundreds or thousands, can form proteins of almost any imaginable size and shape.

In a protein that is 147 amino acids long, it might not seem that one small error—a valine instead of a glutamic acid in the sixth position— would make much difference.[22] But protein shape is everything, and each amino acid contributes to that shape. Assemble a protein polymer incorrectly—substitute a valine for a glutamic acid at Position 6 in the beta globin of red blood cells, for example—and the defective polymer zigs when it should have zagged. The consequence in this example is sickle cell anemia, a devastating blood disease that affects one in every 500 African Americans. The misshaped blood cells carry less oxygen than normal cells have, while their interlocking crescent shapes cause blood cells to be trapped in narrow capillaries. Sufferers of sickle cell disease can experience a host of debilitating symptoms, including anemia, chronic pain, and stroke.

Dozens of other genetic diseases arise from such point mutations: cystic fibrosis causes the production of thick mucous deposits and leads to chronic lung infections; Tay-Sachs disease destroys nerve cells in the spinal cord and brain; other mutations result in a variety of eye disorders that cause color blindness. Errors in the amino acid sequence of proteins, some inherited and some acquired through a lifetime of exposure to physical and chemical hazards, result in a wide range of cancers.

If we live long enough, we all experience the effects of such devastating protein mistakes.

Depolymerization (Weak)

Degradable

Plastics have become a defining miracle of our material age—making possible a spectrum of cheap, versatile products that touch every facet of daily life. The trouble is that we make a lot of plastic, and too much of it is discarded carelessly. From miles of littered shores along the Gulf of Mexico, to giant floating masses in the Pacific, to scraggly bushes festooned with plastic bags in the windswept, dry Sahara desert of Morocco, countless millions of tons of plastic decorate the once-pristine landscape. What to do?

A favored strategy is to design plastics to self-destruct, to gradually decompose once their useful lifetimes are over.[23] The idea is to exploit "depolymerization," a common type of chemical reaction in which a polymer's bonds are disrupted. As the polymer chain or network is fragmented into smaller, unconnected pieces, soluble molecular fragments can simply wash away and return carbon atoms to their natural cycle. New breeds of plastics embrace this opportunity, with a special focus on polymers that can be broken down by hungry microbes.

In some cases, biodegradation of plastic is not an option. After all, you don't want ravenous microbes eating through the PVC plumbing connected to your toilet. But many other disposable plastic products—grocery bags, soda cups and straws, food packaging, diapers, and a thousand other ephemeral items are used once and tossed aside. Of the more than 300 million tons of plastic produced each year, only about 10 percent is recycled. It's best if whatever isn't recycled disappears from the land or ocean as quickly as possible. Of special impact are a new generation of starch-based plastics—polymers that decompose in a matter of months when composted. There's still far too much plastic debris, but thoughtful engineering can help.

Other instances of depolymerization are more problematic, even dangerous. Nylon is a case in point; it degrades when exposed to sunlight over extended periods of time.[24] The process is slow, as the occasional wave of ultraviolet light smacks a bond, severing one of the countless trillions of links in the gradually weakening nylon polymer chain. The process is also all but invisible; you'd probably never notice the atomic-scale changes in a nylon rope's strength.

Phil Rappaport, Harvard graduate student, colleague, and friend, loved rock climbing. He always used the same "lucky rope"—the same rope that had been exposed to years of direct sunlight on beautiful climbing days. When it broke in Wales, on a sunny day in 1974, he fell only 35 feet. But it was an awkward fall, and he died instantly.

Al Dente

Pasta tastes better in Italy.[25] It doesn't hurt that you're enjoying the meal *in* Italy, but mostly it's the distinctive way that Italians make their pasta and then cook it to a pleasingly firm texture—*al dente*, or "to the tooth."

Three factors distinguish the best artisanal pasta. Ingredients provide the first critical difference of the finest Italian pasta. Government controls dictate the use of 100 percent durum wheat flour, known in the United States as semolina. This favored variety of flour has a high content of the proteins known collectively as gluten, which provide the elastic texture of exceptional pasta dough.

The second factor is the careful preparation of the dough—a sequence of tricks learned as much by centuries of trial and error as by any scientific insights. In water, gluten forms a three-dimensional polymer network of proteins that strongly bind to particles of starch (itself a complex polymer of sugar molecules) as the dough is kneaded. A good amount of time is critical; twenty minutes or more must be spent kneading the dough to maximize the contact between the semolina and cold water. Next the pasta is extruded in a press through any of hundreds of different dies to impose the desired shape—conchiglie, farfalle, fusilli, penne, rotini, ziti, and dozens more. Old-fashioned bronze dies are preferable to Teflon because the metal imparts a rougher surface texture to the pasta—one to which sauces cling. Finally, unlike mass-produced pasta, pasta made by hand is dried for a day or two at a modest temperature of 120 degrees Fahrenheit—a temperature low enough that the protein polymers don't break down, while strong bonds form between gluten and starch. The resulting uncooked pasta is dry and brittle, easily packaged and shipped to homes and restaurants.

Cooking is the third key to serving great pasta. The heat of cooking breaks down polymers. That's a good thing if one is dealing with a tough piece of meat or stringy raw vegetables. Marinating can accomplish the same result by "tenderizing" (that is, depolymerizing) food through purely chemical means. But whether vegetables, meat, or pasta, there's a sweet spot between too tough and too mushy. As pasta is gently boiled, it softens by incorporating more water, even as the gluten and starch begin to depolymerize. Great pasta requires just enough cooking to achieve that firm *al dente* texture—but not so much as to turn the pasta to mush.

Brittle (Opus 64, No. 5)[26]

Violinist Fred Shoup is a fixture in the Washington, DC, amateur chamber music scene, a veteran of more than five decades of "jamming." Fred's vast sheet music library betrays the years; his performance copy of Haydn's sixty-eight string quartets is in particularly disreputable condition. At a recent reading of "The Lark," Fred's frayed and yellowing first-violin part, a mid-nineteenth-century edition now re-bound in serviceable gray marbled boards and red cloth spine, literally fell apart. As he quickly turned the first page during a short rest, the embrittled paper fragmented, shattering like a thin plate of glass. Depolymerization had struck again.

This was not the first time, nor would it be the last, that Fred Shoup asked for a pause to carefully align and scotch-tape fragments back together. Paper chips, lost forever, left small angular holes where notes should be. Fred had played the piece often enough to fill in the missing bits from memory.

The four thick volumes of music—first and second violins, viola, and cello—were already ancient and suspect when Fred purchased them a half-century ago in Stuttgart, Germany. Decades of his meticulous penciled-in fingerings and bowings, many erased and re-marked, hadn't helped the already weakened paper. The deterioration of century-old paper, painfully familiar to librarians and collectors of manuscripts and rare books, is an unintended consequence of the

industrial age, when increased demand for cheap paper led to auto-
mated production.

Paper is little more than a glued, intertwined mat of the ubiquitous
biopolymer cellulose, the most abundant biomolecule on Earth. Cellulose fibers, the principal constituent of wood, stems, roots, and leaves,
are polymer chains of hundreds to thousands of glucose molecules,
each a sugar with a ring of six carbon atoms. The strength of paper
rests in its thickness coupled with the lengths of its cellulose molecules.
Industrial papermaking processes of the past 150 years produced thin-
ner, weaker paper with shorter, mechanically pulped cellulose strands.
The use of acids in the washing and binding of mass-produced paper
accelerates the breakdown of cellulose—depolymerization that further
weakens the sheets. Entire libraries of "pulp" magazines, newspapers,
comics, and cheap novels from the nineteenth and twentieth centuries
are turning to dust.

It is difficult, however, not to see the aging of Fred Shoup's music
as a metaphor for us all. We are all depolymerizing: aging skin, hair,
muscle, and bones all weaken as chains of carbon-bearing molecules
break apart.

CODA—Music

CARBON CHEMISTRY pervades our lives. Almost every object we see, every material good we buy, every bite of food we consume, is based on Element 6. Every activity is influenced by carbon—work and sports, sleeping and waking, birthing and dying.

By now you'll have noticed that music is a strong passion of mine, and for that I have carbon to thank. A symphony orchestra—every section, every instrument—sings a song of carbon. The instruments of the string section —violins and violas, cellos and basses—are composed almost entirely of carbon compounds: wooden belly, fingerboard, sound post, pegs, and tailpiece; gut strings, horsehair bow, and plastic chin rest. String instruments also depend on slippery grease for the pegs and sticky rosin for the bow.

The woodwind section? The name gives the game away: wood forms the bodies of oboes, clarinets, and bassoons. Bamboo provides their reeds; cork, the linings of their elegant jointed bodies. Even metal flutes rely on lubricating oil and airtight leather pads for their stunning array of keys.

The percussion section bangs on a riot of carbon—ash drumsticks and calfskin drumheads, teak xylophones and ebony piano keys, castanets and tambourines, woodblocks and claves, maracas and marimbas, conga drums and bongo drums.

Pianos are much the same, with wooden frame, felt-lined ham-

mers, and rubber stops, all hidden in a curvaceous case elegantly fin-
ished with carbon-based paints, stains, and lacquer. And, once upon
a time, the eighty-eight keys of each piano were sheathed in sturdy
veneers of ivory—an expensive embellishment that led to the slaughter
of thousands of elephants per year. One tusk provided enough pieces
for forty-five keyboards; thin plates, three rectangles to a key, were
meticulously cut and then arrayed in the sun for weeks to achieve the
preferred "white" key hue. Today, tough plastics—ivory-colored polymers that simulate the banned carbon-based biomaterial—provide a
benign synthetic substitute.

Ah, you say, but what of the brass family? Surely, trumpets and
horns, trombones and tubas have no need of carbon. Silver-plated
mouthpieces, copper lead pipes, steel valves, brass tubing, U-shaped
tuning slides, and flaring bells are all crafted from solid metal. But fail
to oil your valves or grease your slides, and within a week all you have
is a useless chunk of frozen metal.

Without carbon, all would be silent.

MOVEMENT IV—WATER

Carbon, the Element of Life

Earth, Air, and Fire—enough for a majestic world,
enough for a benign environment, enough for rich stores of
material goods—yet the key essence, Water, is missing.
Carbon is also the element of the living world.

The periodic table has been made. Stars have exploded.
Planets have formed, and a wealth of chemical compounds—
crystal, liquid, gas—have been forged.
Earth is poised to perform its most creative act;
the emergence of carbon-based life is at hand.

Evolution and radiation bring innovation after innovation—
gather the carbon atoms, harness sunlight,
build hard carbon-rich shells, venture onto land.
Life evolves and, in so doing, forever alters Earth's cycle of
carbon as the domains of Earth, Air, Fire, and Water coevolve.

INTRODUCTION—The Primeval Earth

IMAGINE EARTH 4.5 BILLION years ago—an alien, inhospitable world, bombarded by rocks from space, scorched by floods of lava and venting steam, bathed in the young Sun's lethal, unfiltered ultraviolet radiation. No life could have emerged—much less survived—such extreme environmental insults. Yet, in spite of the unforgiving tantrums of infant Earth, the raw ingredients for life were all in place.

Water? Check. The biosphere depends on water. Cells are more than half water by weight, and virtually all origin scenarios demand a watery context. Water is the universal solvent of biology—the milieu in which cells emerge, thrive, and multiply.

Energy? Check. All life-forms require reliable sources of energy, whether the chemical energy of food or the radiant energy of sunlight. And if those tried-and-true sources weren't enough, early Earth also boasted the inexhaustible energy of internal geothermal heat, the pulsed energy of lightning, and the pervasive nuclear energy of radioactive decay.

Carbon? Check. The carbon-based molecules of life rained down on early Earth in a steady barrage of carbonaceous meteorites. Even-richer supplies of life's building blocks emerged from Earth itself—its atmosphere, oceans, and rocks, as our planet became an engine of molecular novelty.

The stage was set. Earth, Air, Fire, and Water were about to organize themselves into something new: Life.

EXPOSITION—Origins of Life

THE STORY OF LIFE'S origins and evolution is an epic tale, best recounted in the language of carbon chemistry. The single greatest transition in Earth's history was the emergence of the biosphere. We know it happened; we are here and driven to understand how it happened. Yet that saga of emergence remains only a vague narrative, largely hidden in the shadows of deep time. A small army of scientists—explorers driven by their curiosity to know what no one yet knows—devote their professional lives to this pursuit. We take on the challenge with no guarantee that a convincing resolution will be found before we die, for this is a journey of discovery that has taken centuries—a quest with far more questions than answers.

Life's Origins—The Five Ws

A plucky reporter will investigate the "five Ws" of any story: who, what, where, when, and why. Add "how" to the list and you have a comprehensive summary of the challenging questions confronting origins-of-life researchers. To these classic questions, we can respond with varying degrees of confidence, though none of these puzzles has been fully solved.

Who and Why?[1]

"Who?" and "why?" in the context of life's origins are questions more suited to philosophers and theologians than to lab-based scientists. Strong, even dogmatic, opinions abound, but science must remain neutral on the why, entwined as it is with the age-old question of life's meaning and purpose. Science relies on independently reproducible observations, experiments, and mathematical logic—an epistemology in sharp contrast to philosophy and theology. Yet science surely informs philosophy; after all, how can we understand the meaning and purpose of the Universe without knowing the ground rules of the cosmic game? Nevertheless, scientists are powerless to tell us why the Cosmos, with all its richly varied living and nonliving bits, exists.

"Who?" is equally unanswerable by the rigorously constrained scientific method, which demands objective, independently verifiable results. Unless we find that life on Earth was purposely seeded by an alien intelligence—a curious concept dubbed "directed panspermia" with its own amusingly speculative literature—the question of who is also beyond the narrow purview of science.

When?

We can be a lot more confident about when life emerged because we have discovered two rigid, bookend-like constraints. On the one hand, the Moon emerged following a spectacular collision between Earth and the smaller hypothetical planet Theia—a cataclysm that disrupted both worlds about 4.5 billion years ago, as suggested by the isotopic ages teased from the oldest lunar crystals.[2] Even if some form of primitive life emerged before that catastrophic Moon-forming event, the impact of Theia created a globe-encircling magma ocean that obliterated any living realm on Earth. Forming the Moon represented a globally sterilizing reset for oceans, atmosphere, and life.

On the other hand, fossil evidence (scrappy fossils from some of Earth's most ancient rock formations in Greenland) reveals that microbial life was well established by about 3.7 billion years ago. Those distinctive stromatolites—mound-like structures with layer upon microscopic layer of minerals deposited by microbes—speak to cellular life that was already highly evolved. We must conclude that life emerged long before those most ancient of fossils.

Exactly when life arose between 4.5 and 3.7 billion years ago remains uncertain. Some experts believe that Earth was habitable, with oceans and an atmosphere, as early as 4.4 billion years ago; a quick emergence of life favors such an early date. Other experts prefer an origin event closer to 3.9 billion years ago, a time after a suspected interval of intense disruption by a swarm of big asteroids and comets. Direct evidence of those globe-rattling collisions has been thoroughly erased from Earth's rapidly resurfaced crust, but the timing and abundance of the so-called Great Bombardment remain vividly etched on the Moon's scarred surface. In any event, we can state with confidence that Earth has been a living world for more than 80 percent of its dynamic history.

Where?

"Where?" in the context of life's origins is intriguing, as it raises questions about unknowable locations that have long been erased from the globe. If, as most of us expect, life emerged on Earth as opposed to some distant world, and if it emerged quickly within a few million years of the catastrophic impact that blasted Earth's outer layers and led to the Moon's formation, then we should consider Earth's cooler polar regions as the most likely place for life's emergence.

Rocks at the poles would have solidified first and were least affected by the immense tidal forces generated by the nearby, newly assembled Moon. Orbiting frenetically, the young Moon cycled through the familiar sequence of lunar phases once every few days. In the millennia after its assembly, the Moon appeared as an immense ball in the sky at less than a tenth of its present distance. Thousand-foot tides must have swept across the perpetually disrupted globe. Only Earth's cooler, stable poles would have been safe from the precocious Moon's pervasive, destructive influence 4.4 billion years ago.

If we assume a more leisurely time frame for life's origins—say, a few hundred million years after the planet formed, when Earth had cooled and the Moon had receded to a safer distance—then the question of where life began on the globe is lost to time. Poles, mid-latitudes, the equator—it hardly makes a difference, and we will never be able to pinpoint the exact GPS coordinates.

But intriguing twists complicate the "Where?" question. Perhaps life originated on some other world and Earth was seeded from afar. This speculative, untested idea comes in at least two distinct flavors. The more "scientific" version looks to a nearby planet, almost certainly Mars, where warm and wet conditions conducive to life may have occurred tens to hundreds of millions of years before Earth became habitable.[3] If life is a cosmic imperative that pops up quickly on any habitable world, then microbes on Mars likely came first. A few of those hardy microscopic bugs, safely nestled in a protective layer of rock, may have hitched rides on Martian meteorites ejected in the violent aftermath of powerful asteroid impacts. Such violent events must have peppered the surface of Mars with regularity.

It may seem counterintuitive, but mathematical models of large impacts reveal that huge chunks of the surface could have been launched into space with relatively little disruption to the rocks or their encased microbial communities. After their relatively short ride to Earth, those microbial hitchhikers could have become the first colonizers, the ancestors to all life today. It might sound far-fetched, but one of NASA's rationales for continued Mars exploration is the search for Earthlike microbes in protected ecosystems beneath today's red, desiccated surface. If discovered, and if those microbes share the biochemical quirks of Earth life, then many of us will conclude that Mars did it first and that we are all descended from Martians.

For the record, a few scholars postulate a more distant origin of life. Astrophysicist Fred Hoyle, who gained fame by discovering the

triple-alpha process that makes carbon in stars, was an outspoken pro-
ponent of a version of panspermia by which virus-bearing comets
brought the first life to Earth.[4] What's more, he argued, they continue
to infect the planet with new viral diseases that rain down from space.
Most scientists view such a scenario as nonsense.

Others speculate on the possibility of life having come from
another star system, perhaps even intelligently designed and purpose-
fully seeded in an act of directed panspermia. Such a hypothesis is, at
least for now, untestable by science. The pseudoscientific idea is also
intellectually flaccid, for it merely transposes the question of life's origins to another place and time. After all, who designed the designers?

What Is Life?[5]

And then there's the maddening "What?" question. If we are to solve
the riddle of life's ancient origins, then we should probably know what
life is. But we don't.

We almost always know life when we see it, but surprisingly,
biologists have so far failed to craft a universally accepted definition.
This lexical shortcoming stems not from any difficulty in recognizing
hopping frogs or swaying birches, but rather from our relative ignorance of cosmic possibilities: we have only one biosphere to study,
only one sampling of "life." If verdant Earth is the only living world
in the Cosmos, then we could easily compile a satisfying laundry list
of chemical idiosyncrasies and physical characteristics unique to our
own biosphere. If we are truly alone in the vastness of space, then our
Earth-based taxonomy would provide a comprehensive definition of
life. We would point to essential chemical ingredients such as carbon
and water, ubiquitous molecular modules like proteins and DNA, distinctive structures including ribosomes and mitochondria, all enclosed
in microscopic cells, the most fundamental common unit of Earth's
diverse biosphere.

On the other hand, if the Universe holds innumerable other living worlds, as many of us who study cosmic history suspect that it does, then it would be presumptuous for us to define life in such a narrow, Earth-centric way. That's why scientists, attempting to distinguish the living from the nonliving realm, resort to lists of more general characteristics and behaviors. All imaginable life-forms—collectively, if not individually—will have the ability to reproduce, to grow, to respond to environmental changes, and to evolve new and novel attributes. NASA, whose long-term mission includes the search for life on other worlds, adds the proviso that life must be a chemical system, composed of interacting atoms and molecules. Accordingly, a computer-based electronic "life-form"—a growing, evolving entity of zeros and ones confined by silicon semiconductors, for example—would be something quite different, requiring a new taxonomy and a different set of organizational rules.

The "What?" question thus encompasses the ambiguity of rigorously defining the essence of life. Scientists approach that taxonomic question with caution and respect, for we have at present only one example of a living world. That state of ignorance might change on any date with a transformative discovery of alien life by one of our planetary probes or direct contact by a more distant alien species. But as of today, we have no scientific basis on which to catalog the range of natural phenomena that might be said to be "alive" (the infinitely creative musings of science fiction writers notwithstanding).

Whatever the still-unproven cosmic diversity of life, efforts to understand the origin (or origins) of life focus on the accessible biology we know best: carbon-based life on Earth. Exploring the ancient transition from a lifeless geochemical world to a planet rich in biochemistry represents one of the most daunting scientific challenges. That ancient transformational leap was too profound to explain with any one theory or to explore in any one set of experiments. Better to divide the story into many comprehensible chapters, each adding a degree of structure and complexity to the evolving world of carbon chemistry.

And that leaves us with the remaining *big* question: "*How* did life arise?"

Life's Origins: The Chemical Ground Rules

When tackling one of nature's great mysteries, it's best to start by examining the ground rules. Three core assumptions frame the study of life's origins. First, most researchers would agree that planets provide all the raw materials—oceans, atmosphere, and a host of rocks and minerals. Many of us also conclude that the origin of life required a sequence of chemical steps, each one adding a degree of complexity and function. And the third and most basic assumption of virtually every origins-of-life investigation is the central role of carbon. Carbon is the cornerstone element of life on Earth today, so most of us in the origins game conclude that it must have been that way from the beginning. But can we be sure?

Making Life: Why Carbon?

Carbon is the element of crystals, of cycles, and of stuff. Carbon, incorporated into myriad solid, liquid, and gaseous forms, plays countless chemical roles that touch every facet of our lives. But what of living organisms, which display structures and functions far more complex than any inanimate material of nature or industry? Which element will provide the vital spark of life?

For a chemical element to be central to life's origins, it must conform to a few basic expectations. Without question, any element essential to life has to be reasonably abundant, widely available in Earth's crust, oceans, and atmosphere. The element has to have the potential to undergo lots of chemical reactions; it can't be so inert that it just sits there doing nothing. On the other hand, life's core element can't be too reactive; it can't burst into flame or explode at the slightest chemical provocation.

And even if an element finds itself at a happy medium of chemical reactivity, in that ideal realm between explosive and dead, it must do more than just one chemical trick. It must be adept at forming sturdy and stable structural membranes and fibers—the bricks and mortar of life. It must be able to store, copy, and interpret information. And that special element, in combination with other ubiquitous elemental construction materials, must find a way to harness energy from combinations of other chemicals or perhaps the Sun's abundant light. Clever combinations of elements must store that energy in convenient chemical form like a battery, and then release controlled pulses of energy whenever and wherever it is needed. The essential element of life has to multitask.

In that restrictive context, consider the many elemental alternatives. The most common elements in the Cosmos are hydrogen and helium, the first and second occupants of the periodic table—the entire upper row—but they would never do as the foundation of a biosphere.

Hydrogen, which can bond strongly to only one other atom at a time, fails the versatility test. Hydrogen is not unimportant, mind you. It helps to shape many of life's molecules through "hydrogen bonding"—a kind of molecular glue—while it plays a vital co-starring role with oxygen in water, the medium of all known life-forms. But Element 1 cannot provide the versatile chemical foundation for life.

Helium, the second element in the periodic table, is of no use whatsoever—impossibly inert, a snooty "noble gas" that refuses to bond to anything, not even to itself.

Scanning across the periodic table, Elements 3 through 5 (lithium, beryllium, and boron) are much too scarce to build a biosphere. At concentrations of a few atoms per million in the crust, and even less in the oceans and atmosphere, they can safely be crossed off the list of prospective life-giving ingredients.

Carbon, Element 6, is the chemical hero of biology; we'll come back to it.

Element 7, nitrogen, is an interesting case. Abundant in the near-surface environment, nitrogen forms about 80 percent of the atmosphere. It bonds with itself in pairs as N_2, an unreactive molecule that comprises most of the gas we breathe. Nitrogen also bonds with many other elements—hydrogen, oxygen, and carbon among them—to form a variety of interesting chemicals of relevance to biochemistry.

Proteins are fabricated from long chains of amino acids, each holding at least one nitrogen atom. The vital genetic molecules DNA and RNA also incorporate nitrogen in their structural units, the "bases" that define the genetic alphabet—A, T, G, and C (adenine, thymine, guanine, and cytosine). But nitrogen, which is 3 electrons shy of the magic number 10, winds up being a little too greedy for electrons: its chemical reactions are a bit too energetic, and the resulting bonds a bit too inflexible to play the multifaceted role of leading actor. As a consequence, we can eliminate nitrogen from the competition.

Why not oxygen? After all, atom for atom, oxygen is the most abundant element in Earth's crust and mantle, representing more than half of the atoms in most rocks and minerals. In the feldspar mineral group, which accounts for as much as 60 percent of the volume of Earth's varied continents and ocean crust, oxygen outnumbers other atoms by an 8-to-5 margin. The ubiquitous pyroxene group features a 3-to-2 mix of oxygen with common metal elements like magnesium, iron, and calcium. And quartz, the most common mineral on the majority of sandy beaches, is SiO_2. It's remarkable to think that when you lie on the beach, soaking up the Sun, two-thirds of what's holding you up are atoms of oxygen. As a consequence, oxygen is, atom for atom, about a thousand times more concentrated in the crust than carbon is.

But oxygen, in spite of its overwhelming abundance, is chemically boring. An isolated oxygen atom starts with only 8 electrons, 2 electrons short of its desires, so it bonds indiscriminately with just about any atom that will make up the deficit. True, oxygen is absolutely essential to all manner of biologically critical chemicals—sugars, bases, amino acids, and of course water. Yet oxygen can't form the requisite chains and rings and branching geometries that are so central to life's intricate architecture. So, we can cross off abundant oxygen from the short list of life's most critical atomic building block.

Fluorine, occupying the periodic table's ninth position, is much worse, being only a single electron shy of the desired complement of 10. Fluorine sucks up electrons voraciously from almost any other element. Reactive fluorine corrodes metal, etches glass, and explodes on contact with water. Breathe a lung full of fluorine gas and you will die horribly, in agony as your lungs blister with chemical burns.

And so it goes. Elements 10 and 18, neon and argon, are inert gases, so we give them no further consideration. Sodium, magnesium, and aluminum (Elements 11 through 13) are too eager to give away electrons; while phosphorus, sulfur, and chlorine (Elements 15 through 17) are too eager to accept them. And as we delve deeper into the periodic table, the elements become less common, and the possibilities for life's core chemistry dwindle.

An exception might be found in the abundant element silicon, which falls in the middle of the periodic table's third row. Silicon is Element 14, occupying the significant position right below carbon. Since elements sharing a column of the periodic table often have similar properties, could silicon be a viable biological backup to carbon? Science fiction writers have seized on this option more than once. I vividly recall an episode from the first season of the classic *Star Trek* TV show—the original one with William Shatner as Captain James T. Kirk and Leonard Nimoy as Mr. Spock—in which the crew of the *Enterprise* discovers a race of intelligent and potentially dangerous silicon-based life-forms shaped like rocks. The concept of the show was fun, especially with the satisfying peaceful resolution as rocks and humans learned to get along. But the mineralogical premise was flawed; silicon is a biological dead end. Silicon at Earth's surface has only one bonding imperative—find four oxygen atoms and make a crystal. Once formed, those silicon-oxygen bonds are too strong and too inflexible to do interesting chemistry. You simply cannot base a biosphere on a single-minded element like silicon.

Keep going, but you will search in vain for another promising elemental option. True, your eye might fall on iron, Element 26, the fourth-most-abundant element in the crust after oxygen, silicon, and magnesium. Why not iron? Iron loves to bond, and it's flexible in its choices. Bond with oxygen? Sure—form red rust with ionic bonds. Bond with sulfur? Of course—make golden, lustrous metallic pyrite (aptly called "fool's gold") with covalent bonds. Iron bonds to arsenic and antimony, to chlorine and fluorine, to nitrogen and phosphorus, even to carbon in a variety of iron carbide minerals. And if no other elements are handy, iron happily bonds with itself in iron metal. Such a diverse bonding portfolio might seem ideal for the core element of life, but iron has a flaw: it readily forms minerals with big crystals, but it shies away from making small molecules. Life demands a huge variety of molecules, with chains and rings and branches and cages—tricks that iron rarely attempts.

And so we are left with carbon, the most versatile, most adaptable, most useful element of all. Carbon is the element of life.

Making Life: What Can Carbon Do?

Short answer: carbon does almost everything. The challenge facing carbon is to forge an astonishing range of molecules to serve the multifaceted functions of life. Shape is one vital attribute. The successful functioning of life's molecules depends to a remarkable degree on their three-dimensional configurations. In some cases, the need for a simple functional form is obvious. Components as diverse as ligaments and tendons, vines and tendrils, spider silk and human hair require strong bonding in one dimension to fashion ropelike, fibrous forms. Carbon accomplishes that feat by linking together in long, strong chain-like polymers.

By contrast, flat layers of carbon-based molecules form the thin, flexible membranes that encapsulate cells, the durable cartilage that lines your versatile joints, and your smooth, supple skin. More complex arrays of molecules serve varied mechanical functions—as tunnel-like molecular passageways into and out of cells, as tiny conveyor belts to move nutrients within cells, as plumbing systems through which fluids flow, and even as submicroscopic molecular motors that propel sperm to their destined rendezvous with a soon-to-be-fertilized egg.

Life also requires a diverse chemical tool kit to accomplish its varied chemical tasks and tricks. Some utilitarian molecules act like tin snips or scissors, processing food by cutting off small consumable fragments from larger particles. Your stomach is loaded with molecules that digest proteins or fats or complex carbohydrates, reducing bulky chunks of food to usable molecular bits. Other molecular tools with exquisitely evolved shapes efficiently staple two smaller target molecules together into a new product, or sort molecules into similar groups, or fold a target molecule into a new useful configuration. Some of these molecular tools incorporate thousands of atoms in three-dimensional shapes of bewildering complexity. More than one Nobel Prize has been awarded for deciphering the structure and function of just one such molecular marvel.

Carbon is the only chemical element that can serve as the backbone for such a diverse array of complex molecules. Its secret resides in its chemical flexibility. As the sixth element, lying halfway between the magic numbers 2 and 10, carbon can achieve a stable environment by adding electrons, by giving them away, or by sharing them in a variety of ways with two or three or four neighboring atoms.

Controlling electrons is the chemical secret to life. Life depends on highly regulated sequences of chemical reactions—intricate processes that take in energy, store that energy, and use the energy to build living tissues. Every one of life's essential chemical reactions involves the rearrangement of atoms and their electrons. Control the movement of atoms and electrons, and you can control the essential processes of life.

Carbon achieves this goal because it can bond directly to dozens of different elements, including itself, to create a wide range of local chemical environments. While most carbon atoms surround themselves with four adjacent atoms, each contributing a single electron to achieve the desired magic number 10, carbon also forms "double bonds," sharing two electrons with another atom, commonly with oxygen or with itself. Double bonds result in carbon atoms with only two or three nearest neighbors, in contrast to the usual four. In special cases, carbon can even form "triple bonds," sharing three electrons with another atom, most commonly with nitrogen or another carbon atom. A triply bonded carbon atom has need for only one additional electron, provided by one additional atomic neighbor. These varied bonding options greatly increase the geometrical diversity of carbon-based molecules.

Some of the resulting configurations—decorating a long chain of carbon atoms with lots of hydrogen atoms, for example—lead to hydrocarbon molecules in which every atom and every electron finds itself in a rather stable, unreactive state. Barring an extreme chemical disruption—being lit on fire in the presence of reactive oxygen, for example—the atoms and electrons of a hydrocarbon molecule stay put. Consequently, long-chain hydrocarbon molecules serve as effective building blocks of protective cell membranes, as well as the principal means of long-term energy storage in fats and oils.

By contrast, the proteins that regulate cells are large carbon-based molecules that rely on moving electrons in exquisitely controlled ways. Their atoms are arranged such that an electron tenuously held by a few atoms—often a cluster containing a metal atom such as iron, nickel, or copper—will readily transfer that electron. A slight change in the molecule's environment will trigger that reaction. One chemical reaction may trigger another and another—a rapid cascade of electron shifts all precisely controlled by the geometries of the carbon-based proteins. Such chains of reactions are essential for building new molecules as cells grow and reproduce.

Carbon facilitates unmatched molecular flexibility because it plays many roles. Carbon accepts electrons, gives away electrons, or shares electrons, thereby bonding to form molecular chains, rings, and branches with single, double, or triple bonds to dozens of different chemical elements. It forms molecules as small as CO, CO_2, and CH_4, but also participates in gigantic molecular structures with literally billions of atoms.

Given this unmatched versatility, it is perhaps not surprising that 90 percent of what chemists do in the lab focuses on carbon. Look at the range of courses taught in any university chemistry or biology department and you will be struck by carbon's disproportionate importance: organic chemistry, polymer chemistry, pharmaceutical chemistry, biochemistry, molecular genetics, agricultural chemistry, food chemistry, and environmental chemistry. Seminars focus on the computer-assisted design of drugs, the intricately folded structures of proteins, carbon-based nanomaterials, the microscopic constitution of soils, and the complex chemistry of wine. All of these topics, and dozens more, center on the chemical richness of carbon.

Strategies: Emergent Steps to Life

A popular game in origins-of-life research is to dream up an "origins scenario"—an elaborate, sweeping, often untestable story of chemical and physical circumstances by which the living world emerged from a lifeless geochemical milieu. Each of these imagined scenarios relies on some previously ignored physical or chemical trick—perhaps a distinctive mineral template, like mica or pyrite; or a surprising physical environment, like a windblown aerosol spray high in the atmosphere; or a sulfide "bubble" near a volcanic vent deep on the ocean floor.

Points (and publicity) are awarded for novelty. British mineralogist Graham Cairns-Smith, a creative scientist, entertaining speaker, and engaging writer, attracted much notice for his "clay world" hypothesis.[6] He speculated that a precocious fragment of clay (the ubiquitous, slippery mineral component of mud) began to self-replicate, transfer information, and evolve—ultimately to serve as a template for the biomolecules of modern biology. Never mind that the mechanisms of the process were only vaguely outlined (and arguably untenable from a crystal chemical perspective); the scenario captured people's imagination, while riffing on the ancient Jewish myth of the golem, a creature born of clay.

Origins-of-life meetings and publications regularly feature such conceptual contributions: the "PAH world," the "mica world," the "borate world"—each a just-so story focused on a novel quirk of nature, each dependent on some special circumstance to facilitate the intractable leap from inanimate chemicals to living planet.

However clever any given scenario might appear at first blush, however appealing the gimmick or passionate the presentation, to me each one seems a bit stifling—a tacit denial of the astonishing richness of natural possibilities. Origins research is, in a sense, analogous to the game Twenty Questions, by which you attempt to identify a mystery individual by asking a sequence of ever-more-narrow yes-or-no questions. The strategic player always begins with the most generic questions—determining whether the mystery person is living or dead, male or female, and so on.

Origins research should be no different. Ask the most general questions first: By what varied reaction pathways does nature synthesize biomolecules? By what mechanisms might those essential ingredients be assembled into functional polymers and membranes? Many origins scenarios, by contrast, are much too limiting—somewhat akin to asking, "Is it Charles Darwin?" as your first question. Of course, scenarios, like inspired guesses, can be original and thought provoking, and once in a great while you might be lucky playing Twenty Questions on a hunch, but it's not a particularly useful or satisfying way to approach the profound scientific question of life's emergence.

There's a better way. The most fundamental approach to answering the origins question is to think of life's emergence as a sequence of chemical steps, each of which added structure and complexity to what would ultimately become Earth's biosphere. First step: you must make the basic modular molecular building blocks—amino acids, lipids, sugars, bases. Next step: those simple molecules must be assembled into functional structures—"macromolecules" that serve roles as membranes and portals, that store and copy information, and that facilitate growth. Final step: the collection of molecules must learn to make copies of itself.

This approach, viewing life's origins as a sequence of emergent steps, has significant advantages over any idiosyncratic scenario, no matter how clever. Each step can be investigated in a targeted and rigorous experimental program. Each step addresses fundamental questions related to carbon chemistry that are important in their own right. And this simple experimental strategy is most likely to mimic the sequential chemical steps that must occur on any carbon-rich planet or moon, anywhere in the Universe.

Step 1: The Emergence of Biomolecules

The first step, one that is as well understood as any aspect of origins research, must have been to make life's essential molecular components. The breakthrough experiment, heralded as the dawn of serious origins-of-life science, took place at the University of Chicago in the early 1950s. Graduate student Stanley Miller, searching for a suitable PhD project, turned to his famed mentor, Harold Urey, for advice.[7]

Twenty years earlier, Urey had been the first to isolate and characterize the heavy isotope of hydrogen, called deuterium—work that won him the 1934 Nobel Prize in Chemistry. During World War II, he joined the Manhattan Project and played a pivotal role in developing the atomic bomb by directing the separation of the fissionable uranium-235 isotope from the much more abundant uranium-238.[8]

After the war, many nuclear scientists shifted away from the kind of applied research that had led to weapons of mass destruction. Harold Urey refocused his energies on the chemical evolution of planet Earth, employing the isotopic record of rocks to infer temperatures of the ancient oceans and compositions of the atmosphere from past geological eras. Among Urey's influential discoveries was the realization that Earth's early atmosphere, dominated by volcanic exhalations before the pervasive influence of life, was radically different from that of today. He posited a mixture of reactive gases, including hydrogen, methane, and ammonia—all potential contributors to prebiotic chemistry. No one knew what chemical reactions such an exotic atmosphere might foster, but Urey speculated on the implications that such a gas mixture might hold for life's origins. Stanley Miller, inspired by one of Urey's lectures on the topic, decided to find out.

Together, Urey and Miller designed an elegant tabletop glass apparatus—an assembly of bulbs and tubes filled with a shallow pool of water and a mixture of gases, gently heated from below and stimulated by electrical spark discharges that mimicked a primitive near-surface environment laced by lightning. The striking results, published in 1953 and trumpeted in headlines around the world, described "a production of amino acids under possible primitive Earth conditions."[9] Miller and Urey had made key molecules of life from the most basic ingredients—water and the kind of gases that might have spewed from volcanic vents on early Earth. It was the seminal contribution in what was to become a growing cottage industry of origins-of-life research.

Did Life Originate at Deep Volcanic Vents?

Intimately connected with the question of how life emerged is one more essential aspect of the seemingly unanswerable "Where?" question. Did life emerge at the sunlit surface or in the dark depths of the ocean? No other origins topic has evoked more heated debate and fueled more controversy.

It's a quirk of human nature that we tend to think in dichotomies. Claude Lévi-Strauss, the twentieth-century French anthropologist and philosopher-author of *The Savage Mind*, characterized such black-and-white perceptions as a throwback to primitive survival mechanisms: quickly recognizing friend versus enemy could prove a life-versus-death choice.[10] Better not to equivocate when faced with mortal danger. Today's news is filled with the ongoing, modern consequences of such a rigid bimodal mind-set, as racism, nationalism, political factions, and religious fundamentalism continue to divide the human population into fragments of "us" versus "them."

It would be gratifying to think that we rational scientists come to our studies with a more nuanced and enlightened worldview, but you need only to glance at the highlights of science history to see that many researchers have fallen into the same trap.[11] More than two centuries ago, the greatest geologists of their day were drawn into intense debates that divided well-meaning researchers into two groups: uniformitarians and catastrophists—the former arguing that all geological processes are gradual and still in play today, the latter invoking brief and cataclysmic events (read: the biblical flood of Noah) as the cause of Earth's geological history. Today it's obvious that the truth lies somewhere in between. A similar rancorous debate raged between Abraham Gottlieb Werner and his Neptunist followers, who favored a watery origin for rocks, and James Hutton's Plutonist disciples, who advocated heat as a principal causative agent of Earth's diverse crustal structures. Once again, both camps had it partially correct.

Stanley Miller's 1953 discovery of the abundant, facile synthesis of amino acids and other bio–building blocks in his tabletop experiment fueled a new false dichotomy. Miller and most of the budding origins-of-life community concluded that a key piece of the biogenesis problem had been solved. Biomolecules formed in an ancient lightning-laced atmosphere. "If God didn't do it in this way," quipped the influential biochemist Leslie Orgel, "then He missed a good bet."[12] The easy, early success proved seductive; the "Millerite" catechism prevailed for more than three decades. A growing army of disciples, trained at Miller's San Diego laboratory, fanned out across the globe to preach the Miller-Urey orthodoxy.

The 1977 discovery of "black smoker" volcanic vents with their rich microbe-hosted ecosystems on the deep, dark ocean floor offered an intriguing alternative origins scenario—one based on the reliable, ubiquitous chemical energy of minerals continuously generated by volcanoes. Rock-powered life was appealing as a plausible and complementary way to make biomolecules—a more benign synthesis pathway that avoided the disruptive, episodic effects of lightning. Many of us (especially mineralogists, whose work suddenly became potentially more relevant) embraced the new idea. But Miller and company fought the hydrothermal origins idea with a vengeance, publishing paper after paper explaining why the "Ventists" were wrong. In a prominent 1992 cover story in the widely popular science magazine *Discover*, Miller denounced the hydrothermal hypothesis as "a real loser."[13] "I don't understand why we even have to discuss it," he complained.

NASA saved the deep origins-of-life hypothesis. The agency's mission to explore other worlds, especially planets and moons that might harbor life, was amplified by the prospect of deep life. After all, if life's origins are restricted to the Miller-Urey model of a warm, wet surface environment periodically interrupted by lightning, then Earth and perhaps early Mars are the only potential sites for life in our solar system. That's a pretty short list for an organization devoted to space exploration. But if a deep, dark, wet volcanic zone can play the starring role, then a host of other worlds become fair game for bio-exploration. Jupiter's ice-covered moons Europa and Ganymede, and maybe even Callisto, show evidence of grand subsurface oceans heated from below—the heat supplied by tidal flexing as the moons orbit the gas giant planet.

Saturn's big moon Titan, though frigid at the surface, has "cryo-volcanoes" with flowing, then freezing, water as the "magma," so Titan might also have deep hydrothermal zones. Even more enticing is Saturn's tiny moon Enceladus.[14] Though only 500 kilometers in diameter, Enceladus boasts a subsurface ocean and hydrothermal vents that blast fountains of water onto its ice-covered surface. Even today's Mars, with presumed warm and wet subsurface environments, begins to look more promising as a home for some sort of primitive underground microbial ecosystem. Given these prospects, however speculative, NASA embraced the hydrothermal origins hypothesis and began funding several laboratories (mine included) in field studies, lab experiments, and theoretical modeling of alternative environments for life.

It has taken more than a quarter century of experiments and debate, but scientists are now showing a rich, plausible prebiotic chemistry associated with deep hydrothermal zones, which must have complemented aboveground synthesis mechanisms. Many researchers are focusing on ubiquitous basalt weathering reactions, by which fresh volcanic flows of basalt are transformed into carbonate and clay minerals while releasing hydrogen—itself a wonderful energy source for life. And, as new evidence for a rich, deep carbon-based chemistry pours in, the misleading Millerite-versus-Ventist debate is quickly entering the annals of science history as just one more example of an unproductive polarization of nature's subtlety.

The lesson to be learned is clear: The imposition of false dichotomies on questions about the natural world serves not only to polarize researchers, but perhaps also to impede scientific progress by ignoring the intricacies of complex systems. Nature is rarely painted in black and white. By shunting aside false and arbitrary divisions, we make more rapid progress toward a nuanced truth.

—

After more than a half-century of research by hundreds of scientists around the world, we have learned that early Earth was an engine of organic synthesis. Essential carbon-based molecules of life—amino acids, sugars, lipids, and more—formed abundantly at the lightning-blasted surface and at deep ocean volcanic vents, in sunlit coves and in warm little ponds. Biomolecules rained down from the heavens, the payload of carbon-rich meteorites, and they formed high in the atmosphere as the Sun's insistent ultraviolet radiation reworked the air.

In the past decade, scientists of the Deep Carbon Observatory have further expanded this impressive inventory, enlisting experiments and theory to highlight the vast potential of Earth and other planets to generate organic molecules in their deep, hot interiors. Researchers in a dozen countries are now synthesizing essential biomolecules and other organic species at extreme mantle temperatures and pressures that most of us thought until recently were inimical to life's essential molecules. The take-home message is plain. Our young planet, and by extension, warm and wet planets and moons across the Cosmos, have been proved adept at making the molecules of life. Perhaps the greatest contribution of the past seven decades of origins research is this unambiguous recognition that the Universe is an engine of biomolecular synthesis.

Step 2: Selection and Concentration

The second step in life's origins presents new and different challenges—not making organic molecules, but winnowing them down. Prebiotic

Earth generated carbon-based molecules in bewildering profusion—hundreds of thousands of different "small" molecules, each with just a few carbon atoms, each available as potential bio–building blocks. Life, by contrast, in spite of its bewildering structural diversity, adopts a more minimalist chemical strategy. Most cells rely on just a few hundred select molecules.

A case in point: Of the thousands of possible varied amino acids, living cells use only twenty different types for most purposes. What's more, most of those twenty amino acids come in at least two versions related by "mirror symmetry"—otherwise identical "left-handed" and "right-handed" variants. Experiments in prebiotic chemistry invariably produce equal amounts of left- and right-handed molecules, but life uses the left-handed amino acids almost exclusively. Similar parsimony applies to life's sugars, almost all of which are of the right-handed variety, as well as to many lipids and the molecular components of DNA and RNA. Consequently, the second challenging step in the path to life's origins is to select just the right subset of molecules and concentrate them, perhaps on mineral surfaces or at the Sun-bathed fringes of a drying tidal pool.

Surfaces are one attractive option—one to which my colleagues and I have paid special note. The vast oceans of ancient Earth were far too dilute for prebiotic molecules to meet and link up with any regularity, but surfaces could have facilitated the mating trick. In some cases, as in the classic oil-in-water scenario, molecules concentrate at the water's surface, thereby forming their own separate layers and globules.

The membranes that enclose cells are a good example; they assemble spontaneously from myriad long, skinny lipid molecules, each with a carbon backbone.[15] One end of each molecule is strongly attracted to water, the other end equally repelled by water. If you submerge lots of these slender, double-ended molecules in water, then attractive and repulsive forces will quickly guide millions of them to line up side by side into a flexible, double-layered, water-filled spherical structure. Water-loving ends of the aligned molecules wind up facing both the rounded outsides and the hollow insides of the spheres, while water-hating ends nestle against each other deep inside the membrane, as far away from water as possible.

Experiments on mixtures of prebiotic molecules have confirmed this membrane-forming mechanism again and again. Gooey molecular mixes scraped from the Miller-Urey apparatus, or extracted from carbon-rich meteorites, or generated from high-temperature synthesis experiments all spontaneously form tiny cell-like structures in water. That part of the origins story—the inevitable emergence of the most primitive cell membranes—seems to be solved.

More problematic is the selection and concentration of the majority of life's molecules—the ones that tend to dissolve in water and don't easily self-organize. How did ancient amino acids find each other to make the first proteins? How did the molecular building blocks of DNA and RNA assemble into the first structures to carry and copy biological information? To solve those riddles, many of us have turned to the mineral kingdom.

Minerals and the Origins of Life

Life's origins depended on a reliable supply of raw materials—the chemical bricks and mortar of cells. But cells depend on just the right chemical ingredients meeting and joining forces—steps that could never occur in the weak primordial broth without a helping hand.

Fortunately, nature invented more than one way to concentrate life's molecules from a dilute ocean. One inevitable mechanism is to have ocean water splash or surge into a shallow pool where it evaporates, thus concentrating the remaining chemicals into a rich organic soup. A century and a half ago, Charles Darwin imagined such a "warm little pond" in a letter to a friend, and the cozy imagery of a benign sunlit venue for life has stuck.[16]

Stanley Miller, perhaps inadvertently, tested a novel variant of this idea by placing a container of organic solution in a low-temperature freezer and leaving it there, evidently forgotten, for three decades.

As the water froze, the remaining tiny pockets of fluid became more and more concentrated into a solution rich in carbon-based molecules, which slowly reacted with each other to produce new organic species. Cycles of freezing and thawing on early Earth might similarly have contributed to an expanding inventory of concentrated bio–building blocks.

In spite of a succession of clever ideas, after decades of futile efforts to coax watery mixtures of life's essential molecules into useful biological structures, many origins-of-life researchers have concluded that the solid foundations provided by rocks and minerals must have played essential roles. The ordered atomic arrays of mineral surfaces likely served many functions in life's origins. Some minerals catalyze the synthesis of key biomolecules—amino acids, sugars, and bases. Other minerals select and concentrate those small molecules, adsorbing them onto surfaces in precise positions and orientations, while protecting them from chemical attacks. Minerals also hold the potential to align and link molecules into functional membranes and polymers.

These ideas are firmly entrenched in the origins-of-life literature, if for no other reason than we can think of few viable alternatives. In the absence of minerals, molecules rarely bump into each other, much less bond in useful ways. In the open ocean, especially near hot seafloor volcanic vents, those fragile molecules are more likely to decompose. Minerals select, concentrate, protect, and link together life's molecular raw materials. Yet in spite of these plausible arguments, few scientists have attempted to perform the exacting experiments necessary to test the roles of minerals in natural conditions.

Experiments on mechanisms by which organic molecules adsorb to mineral surfaces lie at the boundary between biology and geology. They are challenging because they require expertise in at least three fields that rarely overlap. First, you have to be a wiz in aqueous chemistry, because all of the experiments must be conducted in water with precisely controlled conditions of temperature, water composition, and acidity. You also need more than passing knowledge of organic chemistry, especially the complex behavior of amino acids and sugars, which can change their shapes and chemical properties as their watery environment changes. And you'd better be an expert in mineralogy, especially in the nuances of the intricate atomic structures of crystal surfaces.

Few scientists are up to that triple challenge, but one exceptional young researcher, named Charlene Estrada, proved more than equal to the task.[17] From the age of three, when her dad would swing her in the air to the theme of *Star Trek*, she knew she wanted to explore the Universe. Her father, a professor of Mexican American studies, was the first in her family to earn a PhD, so the path to an academic career had been firmly ingrained. "I wanted to be an astronomer, a paleontologist, an archeologist. My favorite toys were magnets, a rubber cast of Jupiter, a pair of binoculars, and (sanitized) chicken bones. Nothing was too strange or scary, as long as it was interesting first," she recalls.

Estrada's decision to become a mineralogist was forged when her family moved to Tucson, Arizona, home of the famed Tucson Gem and Mineral Show—the world's largest. Every year she saved up her allowance just for the chance to build her rock collection during the weeks in January and February when thousands of dealers from around the world came to town, setting up displays in tents, exhibition halls, and hotel rooms across the city. As an undergraduate at the University of Arizona, Estrada gravitated to the mineral lab of Bob Downs, where she became his star student—a budding expert in the field.

I met Charlene Estrada in the summer of 2008, when she joined the Geophysical Lab as an undergraduate intern. She worked in the mineral surfaces lab with Dimitri Sverjensky and me, merging her expertise on crystals with a growing skill in aqueous chemistry. In ten short weeks, she completed an elegant study on the adsorption of amino acids onto rutile, a titanium oxide mineral that is particularly stable in water and thus provided a baseline for all our mineral surface studies.

Estrada's summer was an audition of sorts: her application for graduate studies at Johns Hopkins flew through the selection process, and she was soon working with Sverjensky, immersing herself in geochemistry and surface science. She produced paper after paper, each exploring a different mineral-molecule combination in experiments that reinforced ideas of how minerals might have played key roles in life's origins.

In one remarkable experiment, Estrada added a mixture of five amino acids in equal concentrations to a solution with calcium ions, a typical chemical ingredient in seawater, and brucite, a magnesium-bearing mineral that commonly forms as fresh volcanic rocks react with water and alter on the ocean floor. We thought that all of the amino acids would adsorb similarly onto brucite, but Estrada found that only one of the five molecules, aspartate, easily stuck to the mineral's surface.[18] And in the biggest surprise of all, she discovered that calcium ions gave aspartate molecules a helping hand, greatly enhancing aspartate's adsorption onto brucite. We now realize that such cooperative effects among molecules and ions must be one of the keys to life's origins.

Information

Science is an edifice that builds on itself. Charlene Estrada's advances were but one set of bricks in one corner of that structure, but they pointed the way for others to continue building. The next scientist to join our team was Teresa Fornaro.[19] With a doctorate from the fabled Scuola Normale Superiore of Pisa (adjacent to the Leaning Tower) and an intense research program at the nearby Arcetri Astrophysical Observatory in Florence, Fornaro exemplifies the new breed of astrobiologist—the generation that is likely to unlock the secret of life's origins. Trained with a breadth and depth unmatched in most previous cohorts of young scientists, she is an expert in organic chemistry and mineralogy, planetary science and Earth history, with equal confidence running finicky analytical machines in a surface science laboratory and performing sophisticated quantum mechanical calculations of mineral-molecule interactions.

Teresa Fornaro didn't have to enter the competitive world of scientific research. She could have joined her family business—one of the premier artisanal-pasta factories in Naples, Italy, supplying Michelin-starred restaurants across Europe and beyond with dozens of varieties. But the fast-evolving science of astrobiology has become her overriding passion. She is animated, vibrant when she speaks of her latest findings, and quickly engages you with her confidence and enthusiasm.

Fornaro built on Estrada's work by tackling one of the central mysteries of origins research: the formation of information-rich molecules like DNA and RNA. Many origins-of-life experts focus on RNA as a versatile molecule that, perhaps without peer, has the potential to display several of life's characteristics. RNA can catalyze chemical reactions, speeding up critical biological functions. RNA can also carry information in its four-letter alphabet: A, C, G, and U. And RNA has the potential (though not yet proved in the laboratory) to make exact copies of itself—life's essential property of self-replication. Consequently, the "RNA world" hypothesis is much more than a quirky scenario. For many scientists, it is *the* origins hypothesis to beat. The dilemma is that no one has been able to figure out how to assemble a stable RNA molecule in a plausible prebiotic environment. What's more, the building blocks of RNA are chemically fragile; they tend to decompose in water.

Focusing on brucite and, by implication, on a deep ocean environment that might seem inimical to RNA formation and stability, Fornaro studied how the building blocks of RNA interact with mineral surfaces.[20] Her discovery is profound: brucite selectively adsorbs the building blocks of RNA, while protecting them from the decomposing effects of their watery surroundings. She also discovered that mineral surfaces orient the molecules in a way that might assist in the prebiotic self-organization of RNA.

Admittedly, each experiment is only a tiny piece of the origins story. Decades more work on mineral-molecule interactions is needed. But a growing community of determined scientists is making headway one experiment at a time.

—

These studies reveal an important truth about prebiotic chemistry. Simple experiments with only one or two chemical ingredients, conducted at room temperature and pressure, and using ordinary tap water, are relatively easy to perform, but they also may be misleading when it comes to the processes leading to life's origins. Cooperative and competitive effects among many molecules must have played essential roles in life's emergence. We are coming to the conclusion that the complexity of life arose in large measure from the complexity of the natural geochemical environment. It seems likely that the selection, concentration, and assembly of life's molecular building materials demands intricate three-dimensional processes at the atomic scale— processes that the experiments of Charlene Estrada and Teresa Fornaro have begun to reveal.

The good news is that complex interactions among multiple small biomolecules, chemicals in solution, and mineral surfaces hint at a pathway from geochemistry to the complexity of biochemistry. The bad news is that exploring even a tiny fraction of the potential combinations of molecules, solutions, and minerals is a daunting prospect. The giddy richness of plausible prebiotic environments leads us to imagine literally billions of combinations, each one requiring months of painstaking experiments to explore in any detail.

One thing is sure: we won't be running short of things to study in the lab anytime soon.

Step 3: The Emergence of Self-Replicating Systems

The core mystery of life's origins is how small molecules, individually dead and impotent, arranged themselves into an assembly that could make copies of itself. Individual amino acids, sugars, or lipids, even if meticulously selected and concentrated, are nowhere close to life. Those ingredients must somehow self-organize into systems of ever-greater complexity: they build information-rich polymers, surround themselves with protective, flexible membranes, and invent catalysts that accelerate desirable chemistry, while blocking the production of competing molecules. And then the supreme challenge: they make copies of themselves.

One strategy in the search for such a self-replicating molecular process is to look at the carbon chemistry of modern cells. The key is to identify the most ancient and deeply rooted synthesis pathways—the simplest chemistry that is shared by every living cell. One such primitive biochemical process is the citric acid cycle (also called the Krebs cycle or TCA cycle), familiar to generations of high school biology students as a vital step in the energy flow of cells. The way you probably learned, the cycle starts with the energy-rich, six-carbon molecule citric acid, which undergoes almost a dozen successive steps of fragmentation, each of which releases a bit of energy to drive cellular functions while producing molecules that serve as the starting points for the manufacture of varied essential biochemicals. This carbon-based chemical pathway, employed trillions of times every second of your life to process the food you eat, is built into almost every cell of your body.

A half century ago, biologists discovered that some primitive cells learned to run the citric acid cycle in reverse.[21] Start with the simple two-carbon molecule acetic acid, the basic ingredient of vinegar. React acetic acid with one CO_2 molecule to make three-carbon pyruvic acid. Add another CO_2 to pyruvic acid to make four-carbon oxaloacetic acid. Eight more chemical reactions follow, each adding one small bit at a time—H_2, H_2O, or CO_2—step by step to build larger molecules, up to citric acid, with six carbon atoms.

This "reverse citric acid cycle" is able to make copies of itself. Split citric acid into one molecule of acetic acid and one of oxaloacetic acid, so that one cycle becomes two. Keep going and two cycles become four, four become eight, and so on. As an added bonus, many of the intermediate compounds in the cycle act as the starting points for making other important biomolecules—amino acids to build proteins, sugars to build complex carbohydrates, lipids to build cell membranes, and the building blocks of DNA and RNA.

Given its circular simplicity and biochemical potential, many of us in the origins-of-life game think that the reverse citric acid cycle, or something similar to it, became the first self-replicating system billions of years ago. We equate that chemical innovation to the actual origin of life. Our ongoing experiments replicating primitive Earth environments have reproduced most, though not all, of the cycle's essential chemical steps. We feel tantalizingly close.

Whether the driving factor was a reverse citric acid cycle, or a self-replicating RNA molecule, or some other reproducing system yet to be imagined, an intertwined community of molecules interacted ever so slowly in remarkable new ways. Spontaneously, in a moment of unmatched creative emergence, that molecular community began to make duplicates of itself. Deciphering how that transformation took place—which molecules were involved and in what sequence they reacted with each other—represents the biggest gap in our understanding of life's origins. Our present state of ignorance is not quite as bad as the punch line of a favorite Sidney Harris cartoon, in which a scientist's long and complex mathematical proof scrawled on a blackboard is completed by the phrase "Then a miracle occurs." But gaps in our understanding do remain, and origins researchers continue to search in hopes of finding a simple cycle of molecules that makes copies of itself, while they hotly debate the nature of that first self-replicating system.

In its simplest, pared-down narrative, the transition from inanimate geochemistry to a living world sounds straightforward. Life emerged as a logical sequence of chemical steps, each one adding structure and complexity to a carbon-based molecular network, ultimately resulting in an evolving living world. First came the small molecular building blocks, then functional macromolecules, and finally assemblies of those carbon-based molecules that made copies of themselves. If you attend one of the several origins-of-life conferences and workshops held each year around the world, you will hear a succession of scientists present their latest data and hypotheses with intricate charts and graphs and confident prose. And, it must be proclaimed, we have learned a lot about life's emergence billions of years ago. But so much more remains unknown, hidden, a source of mystery.

And that is why origins-of-life research is so much fun!

Second Genesis: Life on Other Worlds[22]

Life on Earth emerged billions of years ago, as the geosphere transformed into the biosphere. In that time, life on Earth has evolved, increasing its dominion from the deep sea to land and sky, while expanding its repertoire of survival strategies.

What we do not know is whether this epic story is unique to our planet, or has been recapitulated countless times on innumerable worlds across the galaxy. Is life one story or many—a cosmic imperative or a fluke? Philosophers are not shy about debating the point. Jacques Monod, Nobel Prize–winning French biologist, took a rather pessimistic view. In his 1970 classic book *Chance and Necessity*, Monod concludes, "The universe is not pregnant with life nor the biosphere with man. . . . Man at last knows that he is alone in the unfeeling immensity of the Universe, out of which he emerged only by chance."[23]

Many scientists, including virtually all of us who spend time in origins-of-life research, are uncomfortable with that negative assessment. Otherwise, we would be wasting our time in the lab. Belgian biologist Ernest Schoffeniels spoke for many of us in *Anti-chance*, his 1976 rebuttal of Monod, when he considered an alternative philosophical stance: that "the origin of life and evolution were necessary because of conditions on Earth and the existing properties of the elements."[24] It's a fun debate, but nobody really knows the answer.

Chance or necessity? Are profound cosmic transitions really reducible to this stark choice or, as I tend to think, is such a polarized dichotomy false? In a Universe rich with emergent, evolving systems—stars and planets, isotopes and elements, life and the conscious brain, society and culture—is it not natural to ask which phenomena are inevitable as opposed to those that are frozen accidents? In no domain of inquiry is this question more relevant, yet more challenging, than life's ancient origins. Is the transition from geochemistry to biochemistry an intrinsic characteristic of Earthlike planets, or is life rare in the Cosmos?

There aren't a lot of hard data to back either view. In the entire expanse of the Cosmos, we have evidence for only one living world and only one origin of life. We will never be sure whether life is a cosmic imperative unless we find a "second genesis"—a second independent origin of life. If we do, then the rule of zero, one, many will kick in. In short, "zero, one, many" means either that a natural phenomenon never happens (time running backward, for example), or it happens exactly once (a "singularity" like the Big Bang), or it has happened more times than we can count (perhaps the origin of life).

On the Combinatorial Richness of Earthlike Planets[25]

It's only natural when thinking about the probabilities of life's origins to imagine running experiments within the time and space constraints of a research laboratory. The three- or four-year span of a typical PhD thesis exploration frames many research projects. But chemical reactions that are unlikely to be observed at such restricted scales of laboratory experiments may, nevertheless, be inevitable in the grand context of planets.

If we conceive of life's origin as a sequence of chemical reactions (and my bias is for chemical reactions promoted by mineral surfaces), then we should ask how many such reactions might have occurred on prebiotic Earth. The answer is nothing short of astronomical. Earthlike planets possess fine-grained clays, volcanic ash deposits, weathering zones, and other exposed minerals with immense total surface areas that are many millions of times greater than the idealized smooth surface of a planet-sized sphere. Those mineral surfaces promote molecular reactions for hundreds of millions of years.

By contrast, individual chemical reactions occur at the minute scale of molecules in a matter of seconds. Even a modest-sized planet can combine and shuffle organic molecules over and over again, effec-

tively attempting different chemical reactions more than a trillion trillion trillion trillion times.

The implications for origins-of-life research are clear. Experiments
requiring exacting conditions or unusual juxtapositions of several reac-
tant molecules might never occur in the restricted conditions of controlled lab experiments, yet they may be inevitable, given planetary
scales of space and time. If life's origin requires one or more of those
unlikely chemical reactions, then we may be hard pressed ever to solve
the problem.

No need to despair. Strategies exist to increase the likelihood of
observing improbable chemical reactions in the laboratory. We can
work backward from modern biochemistry to focus on key molecular species and their products. New approaches in combinatorial
chemistry, coupled with computational chemistry, hold the promise
of quickly narrowing the search. Chemical and physical intuition will
also continue to play central roles in origins research. Nevertheless,
if elucidation of the origins of life depends on a finicky reaction that
occurs on an Earthlike planet only once in every trillion trillion trillion interactions, then a detailed understanding of origins chemistry
may be beyond current laboratory capabilities, even while life's origin
is a deterministic feature of any warm, wet terrestrial world.

Whether life is a widespread cosmic imperative or a unique frozen
accident limited to Earth, we do have one rich and wonderful living world to explore. Carbon-based life emerged 4 billion years ago,
evolving variation after variation on the cellular theme, each variation
adding new wrinkles to Earth's dynamic carbon cycle.

DEVELOPMENT—Life Evolving (Theme and Variations)

Theme: Life Evolving

If you could journey back in time 4 billion years to a young Earth, you would discover a planet taking its first halting steps as a living world. If you took that journey through deep time and walked on Earth's few emerging fragments of land or swam in the globe-encircling oceans, you could be forgiven for missing the subtle signs of life. The sparse population of Earth's earliest microscopic cells found deep, dark refuge in the oceans, clinging to rocks, largely hidden from the relentless solar blast.

The earliest stages of Earth's chemical evolution—the emergent steps leading to life—involved trying and rejecting vast numbers of molecular arrangements until one exceptional consortium of molecules began to copy itself. From that instant on, as new copies flooded the environment, evolution by the Darwinian process of natural selection took over. Inevitably, some of those copies incorporated mutations: oxygen atoms substituted for sulfur, and vice versa; clusters of carbon atoms attached as newly sprouted branches and rings; and other spontaneous variations modified the configurations of atoms, their folds, twists, and kinks. Most of these random mutations had little effect; others were lethal—literal dead ends. But once in a while, a molecular variation worked a little better, facilitating more efficient copying, or a cycle of molecules less prone to degradation, or a molecule capable of surviving in more extreme conditions of temperature or acidity.

Evolution is the theme—evolution by the powerful process of natural selection. As Charles Darwin realized more than a century and a half ago, living systems evolve because of three attributes of life that we all observe on a daily basis.[26] Darwin's first premise is that individuals of every species exhibit variations: no two oak trees or mushrooms or people are exactly the same. We now know that those variations go far beyond the superficial sizes and shapes of creatures great and small; they extend to fundamental differences in thousands of proteins, the essential molecules of life.

The second attribute of all life-forms is that more individuals are born than ultimately survive to adulthood. This truth, amply displayed by the profligate production of acorns every fall and pollen every spring, was forever etched in my mind during a memorable third-grade show-and-tell. The scene was Garnett Elementary School in Fairview Park, Ohio. I usually brought in an odd fossil or mineral collected at nearby Rocky River Park, but this time I had found a strange cocoon-like mass adhering to a twig. Show-and-tell wasn't until late morning, so I put the twig and cocoon in my slant-topped desk and forgot about it.

When I opened the desk a couple hours later, my first impression was that some kind of dark liquid had spilled on my books. Closer inspection revealed thousands upon thousands of tiny scrambling baby praying mantises. They were everywhere—a seething brown mass on my books, pencils, and now inedible lunch. Seeing the daylight, they swarmed up and over the metal edges of the desk, dropping onto the floor and my pant leg. I screamed. Then others came over to look and they began to scream, and soon the classroom was in utter chaos. A brave janitor with a Power Vac came to our rescue, but it took a while for everyone to settle down. (After the excitement, the consensus was "best show-and-tell ever!") In any case, it was an awesome display of Darwin's second premise: more individuals are born than can survive.

Darwin's third key attribute of evolving life is equally obvious: individuals with beneficial traits that enhance chances of survival are more likely to pass those desirable traits to the next generation. If a plant is better able to withstand a cold snap or drought, it will multiply. If an animal possesses better camouflage or superior intellect, it is more likely to have babies. Desirable traits ultimately win.

Given these three fundamental attributes of the living world, all you need is many generations (read "a lot of time") for life to evolve to new forms with enhanced abilities to survive and reproduce. That's the essence of evolution by natural selection.

Once the first cell took hold and began to divide, increasingly intricate feedback loops between cells and their environments led to novelty on land, at sea, and in the air. Six contrasting variations of life's evolution on Earth exemplify the interplay between geological settings and biological novelty. The first life-forms, single cells too small to see with any but the most powerful microscopes, relied almost exclusively on the chemical energy of rocks. It took perhaps a billion years for the second variation to emerge, when more advanced cells began to harvest sunlight as a new energy source. A third variation, starting roughly 575 million years ago, saw the emergence of multicellular life—a new strategy for survival.

Soon thereafter, a biological arms race ushered in yet another variation, the fourth, as mineralized teeth and claws were pitted against armored shells and bones. A fifth variation occurred when plants and animals ventured onto land to create what most of us now think of as Earth's characteristic green landscape. And a sixth variation on the theme of evolution—the most recent variation—finds humans playing a dominant role in Earth's changeable biosphere. Each variation saw a transition to a new strategy for harnessing nutrients and sustaining life. Each evolutionary variation was, in effect, an exploration of energy, and each changed the way that carbon cycles among reservoirs near Earth's changeable surface.

Variation 1: Microbes Eat Minerals[27]

In the story of life, carbon is the drama's central atomic player, but energy drives the action. We are confident that origins chemistry centers on carbon because carbon lies at the core of all biomolecules today—and there is no viable alternative. The question of life's energy source is much subtler, the range of plausible answers more varied. Most life on Earth today ultimately derives its energy, either directly or indirectly, from the Sun through photosynthesis. But gathering sunlight and converting it into chemical structures is an intricate and correspondingly more recent development, requiring the evolution of complex layers of cellular innovations. The most primitive single-celled organisms on Earth today employ a much simpler and presumably more ancient energy solution: they "eat" minerals.

Coming to grips with the concept of minerals as food requires a rather unconventional perspective, which is why the idea appealed to geobiologist Paul Falkowski.[28] A baby boomer, born and raised in New York City, Falkowski recalls growing up "at the edge of Harlem" in the 1950s and 1960s. The working-class Falkowski family struggled to make ends meet. Neither of Paul's parents was particularly interested in science, but they encouraged their son's blossoming fascination with the natural world by giving him a microscope (an extravagant birthday gift when he was nine years old) and taking regular trips to the glorious American Museum of Natural History (where so many of us in science first caught the itch). With the additional encouragement of a young couple living in his apartment building—both of them biology graduate students at Columbia University—Falkowski's lifelong love of cultivating and studying tropical fish and their complex confined ecosystems was born. Even today, his office and labs at Rutgers boast spectacular aquaria with all manner of colorful fish and corals.

Falkowski stayed close to home for his education, attending Brooklyn Technical High School and then the City College of New York. After a brief fling studying philosophy and logic, and acing the litany of required physics, math, and chemistry offerings, he discovered his true calling: oceanography. City College had begun a program sampling water and microorganisms in the Hudson River and New York Harbor on a 90-foot catamaran, the *Atlantic Twin*, and Falkowski, then a senior, volunteered as often as possible.

On paper, Paul Falkowski is a card-carrying oceanographer, with a PhD from the University of British Columbia and many months at sea to his credit, including trips to West Antarctica and the Sargasso Sea. Yet he has enjoyed an idiosyncratic path, working for years at Brookhaven National Laboratory on Long Island, a renowned center for physics research, and as a professor of geology at Rutgers University in New Jersey. A voracious reader, he sees connections in nature that others have missed.

One of Falkowski's most profound insights is that Earth acts like an immense electrical circuit, with life playing a role as fundamental as the light bulb you're (most likely) using to read this book.[29] Here's what he means: Every electrical circuit has three essential components. First, there has to be a reliable source of electrons; that's because electricity is nothing more than the flow of electrons. Second, there must be some kind of electrical conductor through which the electrons flow. And third, there needs to be a place to store all those electrons once they've moved. Looking at the planet on the grandest scale—oceans, atmosphere, rocks, and life—Falkowski recognized that all the components are there. Electrons come to Earth's surface via volcanoes, especially ocean floor volcanoes that carry electron-rich iron atoms from the deep interior. In this sense the rocks are equivalent to the negative terminal of a battery.

The oceans are the "wires" in Falkowski's global circuitry; they conduct electrons away from the electron-rich rocks. Ultimately, those electrons wind up in the oxygen-rich atmosphere, which is analogous to the positive terminal of a battery. Fresh, new volcanic rocks that erupt onto the ocean floor represent a source of electrical potential energy—energy that's just waiting to be used.

Enter the mineral-eating microbes. More accurately, these

microbes exploit the chemical imbalance that arises when some minerals have too many or too few electrons compared to their surroundings. Just as in a battery, huge numbers of electrons are ready to flow, for example, from iron-rich olivine in rocks erupted from Earth's interior to the surface environment. Microbes interpose themselves, in effect plugging themselves into that flow of electrons and short-circuiting their surroundings. That electrical energy is a bacterium's version of a free lunch. In the process, the olivine is gradually consumed, and new minerals form in its place.

Earth's earliest life eked out a living by finding minerals that were out of chemical balance with their surroundings. The microbes grew on mineral surfaces, acting as catalysts, speeding up chemical reactions that replaced older minerals with new species—new minerals that likely would have formed anyway, albeit much more slowly. Virtually all traces of the earliest cells have disappeared from the fossil record. The only tangible results of their ancient microbial banquets are distinctive, typically layered deposits of mineral products.

The secret to these life-sustaining mineral-microbe interactions lies in the ability of many metal elements to exist in chemical varieties with differing numbers of electrons. Iron provides the most prevalent example. Most of the iron atoms that erupt in volcanic lavas are in the +2 state, having given away two electrons to surrounding atoms. But in the presence of oxygen or some other electron-hungry element, +2 iron can give away one more electron, release a tiny jolt of energy, and remain quite happily in the +3 state.

Primitive microbes, looking for reliable energy sources, learned to catalyze iron's electron shift from +2 to +3 and, in the process, precipitate layers of rusty, red iron oxides onto the ocean floor. Indeed, the world's largest mines of iron ores, as well as those of manganese, uranium, and other valuable elements, were formed atom by atom from the actions of countless mineral-consuming microbes.

Throughout the subsequent billions of years, life has expanded its repertoire of energy-gathering strategies, but this most ancient process of exploiting rocks for energy has persisted, becoming an integral part of the global electrical circuit, accelerating and expanding the electron flow of planet Earth.

The Unexpected Abundance of Deep Subsurface Life[30]

Here's a remarkable fact. Drill a hole a mile deep almost anywhere on Earth—in the desert or forest, on land or at sea, near the equator or above the Arctic Circle—and the chance is almost 100 percent that you will find microbial life. There won't be a lot of cells, and they won't be fancy—just little spheres or rods barely visible in a powerful microscope—but you will find living cells. This hidden biosphere, composed almost entirely of microbial descendants of the first mineral-consuming life-forms, provides compelling evidence for Earth's most ancient energy-harvesting strategy.

In no domain is the coevolution of the geosphere and biosphere more obvious than in the deep subsurface realm, where rocks and deeply circulating water are the only sources of energy and nutrients. The study of the deep microbial biosphere has come into its own in the past decade with the Census of Deep Life, a centerpiece of the Deep Carbon Observatory.[31] The census documents subsurface microbial communities from around the globe, primarily from drill cores and deep mines.

Drilling for microbes is an odd concept, but it has become a passion for a select cadre of scientists from around the world. They venture to remote sites on land—Oman, central China, the mountains of Scandinavia, and the deserts of Africa—bringing up miles of cylindrical rock cores. They have penetrated the muddy bottoms of every ocean on Earth, as well as dozens of lakes, from the equatorial regions of Africa and South America to above the Arctic Circle, to retrieve and characterize the sparse populations of underground cells. Exquisite care must be taken to avoid surface contamination, for a minuscule droplet of surface water could swamp any biological signal from the deep.

University of Rhode Island oceanographer Steven D'Hondt has discovered more than his fair share of deep microbes.[32] Sporting a shock of unruly hair and sharing his spontaneous broad smile, D'Hondt gives the first impression of a person having a lot of fun doing science. Like so many of us, he caught the science bug early. "My interest in geology and paleontology was set when my parents gave me a Porter-Spear mineralogy kit for my seventh birthday," he recalls. "It included a manual of rocks and minerals; several diagnostic tools—a streak plate, magnifying lens, and alcohol lamp; and a collection of mineral unknowns to identify." The alcohol lamp would probably fail twenty-first-century safety requirements, but, Steve laughs, "I managed to not burn anything of consequence."

Not everyone was sympathetic to D'Hondt's elementary school interests. "For the next few years, I had to sneak geology, biology, and astronomy textbooks in and out of my school library because the librarian felt they weren't age appropriate." Undeterred, he pursued science with a passion, receiving his undergraduate degree from Stanford and PhD from Princeton.

Steve D'Hondt's professional career has centered on the complex interplay between life and Earth over geological time. Focusing on fossil and chemical records of the past 100 million years, he became increasingly aware that Earth's oceans are continuously altered by the metabolic activity of microbes living within ocean floor sediments. "I realized that the study of deep life offered an exceptional and novel opportunity to understand the limits of life and the influences of microbes on Earth," he reminisces.

Anticipating the DCO by almost a decade, D'Hondt served as co–chief scientist aboard the drilling vessel *JOIDES Resolution* in the winter of 2002. The drilling expedition in the eastern Pacific was the first to focus primarily on subsurface life, and it proved beyond doubt that deep life is a diverse, abundant, and largely overlooked aspect of Earth's biosphere.

Subsequent research has led to significant insights beyond the star-tling realization that Earth's biosphere extends deep beneath the sur-face both on land and at sea. We now understand that subsurface life plays key roles in the chemical evolution of Earth's crust, recycling nutrients while breaking down rocks at a global scale.

The deep microbial world challenges conventional ideas about ecosystems. D'Hondt calls it the "deep zombie-sphere" because subsurface microbes do almost nothing—barely alive, seldom moving, almost never reproducing, and existing at unlikely scales of space and time. For one thing, deep life is astonishingly slow, with cellular reproduction rates perhaps as low as once every thousand years. The average microbial community in many subsurface ecosystems persists for millions of years doing almost nothing, surviving on energy fluxes that are orders of magnitude smaller than those of the surface world. The deepest life, more than 2 kilometers down, may have a population density of only one microbe per cubic centimeter (roughly the volume of a sugar cube). That's analogous to a human population with individuals separated by an average distance of 400 miles.

The deep biosphere has been called the "new Darwinian Galápagos," because deep isolated microbial populations, like the diverse "Darwin finches" on isolated islands of the Galápagos Islands, provide a natural laboratory for studying microbial evolution, diversification, and distribution. Deep life is slow and sparse, yet integrated over all the oceans and continents of Earth to depths of a mile or more; the total subsurface biomass is staggering.

When thinking of this hidden, mysterious, deep biosphere, you might ask, How much life is hidden away? How much carbon is buried in subsurface cells? And how deep does the biosphere extend? The Census of Deep Life now holds records of more than 1,200 subsurface localities, some as deep as 2 miles, with data on the diversity and lifestyles of these hidden denizens. As our understanding of the deep microbial biosphere expands, a few findings stand out. For one thing, the deep biosphere is remarkably extensive. A few years ago, when most subsurface samples came from the microbe-rich oceanic sediments near the margins of continents, it appeared that the total subsurface biosphere might rival that of all surface life—all of the trees and grass and ants and whales combined. Even at depths of more than a half mile, the near-shore sediments typically hold more than a million microbes per cubic inch. Integrated over Earth's vast volume of shallow subsurface sediments, that's a lot of microbes.

Recent measurements of more distant localities, where ocean sediments are far from the nutrient-rich coastlines, reveal far fewer microbes per cubic inch in the deep ocean. Nevertheless, revised estimates of the hidden deep microbial population point to 6×10^{29} cells, constituting 10 to 20 percent of Earth's total biomass—a small but endlessly fascinating portion of Earth's carbon cycle.

Survival of life at that depth poses at least three daunting challenges: pressure, temperature, and energy. It turns out that pressure is not much of a constraint. About twenty years ago, just to see what would happen, my colleagues and I squeezed a culture of the familiar gut microbe *E. coli*.[33] We used the same kind of diamond-anvil cell that is employed in studies of Earth's deep carbon mineralogy, except we filled the sample chamber with water, nutrients, and living microbes. We had intended to raise our screw-tightened pressure apparatus to about 2,000 atmospheres, or roughly twice the pressure of the deepest ocean trench. But some overzealous screw turning caused the pressure to soar above 10,000 atmospheres, equivalent to the crushing pressure 30 miles deep in the crust (sort of an "oops!" moment in the world of high-pressure experiments). Amazingly, some of the microbes survived. From these and subsequent (more controlled) experiments, we concluded that Earth's microbial ecosystems are not limited by pressure.

Such extreme pressures do raise questions. How can the molecules of life, with functions that are so exquisitely sensitive to exacting molecular shapes, not be crushed at pressure? In some instances, clever carbon chemistry is the key. Cell membranes at low pressure form from arrays of molecules with straight carbon backbones, which easily line up side by side like uncooked spaghetti in a box. They pack efficiently but with large enough gaps that essential food molecules can cross the membrane.

At extreme pressure, that arrangement would become too dense; life-giving nutrients wouldn't be able to flow through. The high-pressure versions of membranes therefore adopt kinked carbon backbones, each crooked molecule having several doglegs. When lined up side by side, they adapt to high pressure by flexing like springs, providing pathways for nutrients while not crushing together.

Temperature is a different story. You might think that the boiling point of water, 212 degrees Fahrenheit, would be life's absolute limit. Yet pressure raises liquid water's stability; water temperatures at the deepest seafloor volcanic vents can exceed 550 degrees Fahrenheit. A more fundamental limit appears to be the breakdown of vital proteins, some of which decompose at about 260 degrees. That's hot enough to give you terrible blistered burns, but a few hardy microbial species can survive that extreme heat. For now, 260 degrees Fahrenheit is widely accepted as a limit to cellular life as we know it.

Earth's temperature and pressure profiles are linked: the deeper you go, the hotter it gets. In some "hot spots," like the hydrothermal zones of Yellowstone or Iceland, you need to descend only a few meters to exceed that temperature limit of life. But in cooler continental zones, far from any volcanic activity, temperature increases by less than 20 degrees Fahrenheit per mile of depth in the crust. Consequently, it seems quite plausible that some microbes live more than 10 miles beneath the surface, though probing to such a depth and returning a rock sample is beyond any current drilling technology.

The third challenge to deep microbial life is finding a reliable energy source. Many deep microbes are confined to tiny pockets of water, sometimes isolated for millions of years. Any chemical energy in the mineral grains lining those fluid-filled cavities has long since been exhausted, but recent studies have revealed another quite unexpected energy source: radioactivity. Every rock has a trace of radioactive uranium, perhaps one uranium atom in every million atoms. The energetic decay of uranium is exceedingly slow, with about half of the atoms spontaneously transforming and emitting destructive alpha particles every 4.5 billion years. But rocks host so many uranium atoms

that a slow and steady stream of alpha particles pervades the subsurface realm. When an alpha particle strikes water, it can split H_2O
into hydrogen plus oxygen—a perfect food for microbes. It's not a
very large energy source, but apparently it's enough to keep some tiny
microbial communities happy for eons.

As a mineralogist, I'm drawn to the idea that the story of life is
inexorably tied to the mineral kingdom. Rocks and minerals may have
provided the energetic starting point for life, but another even more
promising and reliable energy source beckoned. And so life learned to
live off the light of the Sun.

Variation 2: Life Learns to Use Sunlight for Energy[34]

For a billion years or more, Earth's primitive water-bound microbial life, including both cells living at the surface and those more
deeply buried, played an insignificant role in the cycling of carbon.
Earth's total biomass was meager, confined to small, sparse coatings of
microbes whose distributions were dictated primarily by the chemical
energy of fresh volcanic rocks exposed to ocean water. That situation
was about to change, as life discovered ways to exploit a much greater
source of energy: the light of the Sun.

Photosynthesis is a remarkable biological innovation. At its core, the
photosynthesis we know today takes readily available commodities—
simple molecules of water and carbon dioxide plus the energy of
sunlight—and manufactures the entire range of molecular products
necessary for life (along with the vital gas oxygen). Complex details
aside, this process represented a fundamentally new and efficient pathway for carbon cycling.

Photosynthesis relies on a fortuitous characteristic of waves of
light, or "photons," streaming from the Sun: Photons carry energy;
the shorter the wavelength, the greater the energy. What's more, that
energy can be transferred from photons to atoms by the process of
absorption. But, like Goldilocks and the three bowls of porridge, there's
an energetic sweet spot—not too hot and not too cold. To induce the

critical chemical reactions of biology, you need to absorb just the right amount of energy to control electron shuffling among atoms.

Atoms easily absorb infrared photons, with wavelengths longer (and thus of lower energy) than those of visible light. Infrared waves cause atoms to wiggle a bit faster—what we feel as heat energy. When you feel the warmth of the Sun or a glowing fire, your skin is absorbing infrared radiation; the molecules of your skin heat up. The effect is exaggerated if an object is black, as you've undoubtedly experienced when you walked barefoot on an asphalt surface on a sunny summer day. Yet only the most energetic of infrared photons—those with wavelengths close to what we see as red light—possess enough energy to shift electrons among atoms and thus drive biological reactions.[35]

At the other end of the light spectrum, ultraviolet radiation has shorter wavelengths, and thus greater energies, than visible light has. These potentially dangerous photons have enough energy to knock some electrons completely off of atoms—a process, called "ionization," that can disrupt atomic bonds and fragment critical molecules. If you've ever suffered severe sunburn, when broken molecules cause skin cells to die, you've experienced ionizing ultraviolet radiation. Damaging UV photons possess too much energy for most biological needs.

Photons of visible light, especially those near the less energetic red end of the spectrum, find themselves at a happy medium. As clusters of atoms like the green plant pigment chlorophyll absorb photons of red light, their electrons jump into excited states. Those electrons can hop from one atom to another, forging new chemical bonds. Photosynthetic microbes take advantage of this fortuitous attribute of visible and nearest-infrared photons to power biology.

Smelly Life

The earliest versions of photosynthesis, invented more than 3 billion years ago, differ from the textbook story you may have learned—the one that generates Earth's oxygen-rich atmosphere. Instead, the first cells to harness the Sun's energy employed other chemicals, like the smelly, toxic gas hydrogen sulfide, H_2S, a common product of volcanoes, to drive biology. These "green sulfur bacteria" employed a light-gathering process called "Photosystem I" to absorb the Sun's red wavelengths and shuffle electrons.

In the first step of this process, the green pigment releases an electron, which then moves to other atoms. Paul Falkowski introduces a clever metaphor in *Life's Engines*, his engaging book on microbial evolution. Shuttling electrons among atoms, he notes, is something like moving people between subway stops at rush hour. Think of yourself as a negatively charged electron waiting on that green pigment. Two negative electrical charges repel each other, so you're not going to voluntarily hop from your comfortable molecular platform onto a subway car filled with other negative electrons. But if a uniformed station manager gives you a shove (as they do in some countries!), then you might find yourself jammed into the car, at least for a stop or two, before you emerge onto a less crowded molecular platform.

In the same way, a photon of red light can provide the energetic push to temporarily move a negative electron from the pigment to other atoms. With the electron gone, the pigment finds itself positively charged and needing another electron. In photosynthesis, that electron can be shuttled from a metal atom, which in turn can steal yet another electron from hydrogen sulfide in a cascade of chemical reactions that splits hydrogen sulfide into hydrogen plus sulfur. The end result is that those reactive products provide a terrific chemical fuel for life.

Not to be left out, other microbes evolved a quite different biochemical pathway, "Photosystem II," to take advantage of the Sun's free lunch. These microbes, including the so-called purple bacteria, absorb slightly more energetic photons to shift and shuttle electrons, but the end result is similar. Electrons move in a cascade of reactions that ultimately splits sulfide and generates the fuel of life.

Though they emerged more than 3 billion years ago, green and purple bacteria still survive in isolated, submerged communities. Search for them in deep, stagnant waters where sunlight penetrates into zones with absolutely no oxygen—a lethal toxin for these primitive microbes. Yet the most lasting legacy of these two types of primitive cells lies not in their sparse distribution on Earth today, but rather in a dramatic evolutionary innovation that linked their two distinct light-gathering strategies into the modern two-stage version of photosynthesis.

Water Power

Sulfides serve as an adequate chemical fuel for green and purple bacteria. Starting at least 3.5 billion years ago, resilient clans of microbes have employed sulfides plus single photons of red light to harvest energy from the Sun. Yet sulfides are not ubiquitous at Earth's surface, nor is the energy of sulfur and hydrogen the best that Earth has to offer.

Water is a much better option. Water—hydrogen oxide—is not only vastly more abundant than hydrogen sulfide, but it also has the potential to provide biological fuel, in the form of oxygen and hydrogen, that yields much more energy. Water represents the ultimate fuel for life on Earth, but there's a challenge: It takes a lot more energy to split water than to split hydrogen sulfide. A single photon won't do the trick. It took a billion years of trial and error, but eventually a lucky microbe discovered the two-step process of oxygenic photosynthesis—a coordinated combination of Photosystems I *and* II. That sequence of photon absorptions provides the extra energy boost required to split water into hydrogen and oxygen.

Living cells thrive on the cascades of chemical reactions that follow the splitting of water. Most important, molecules of hydrogen plus carbon dioxide form the sugar glucose, with waste oxygen left over. Glucose molecules bond together in sturdy chains of "cellulose," the major constituent of green biomass and consequently the most abundant biomolecule on Earth. Stems and leaves, roots and branches, grass and tree trunks—half of modern Earth's biomass is cellulose. The consequences of rampant cellulose production have been profound. Billions of years ago, as photosynthetic cells soaked up sunlight, carbon dioxide in the atmosphere became the feedstock of living cells. Air and water gradually converted into masses of green algae that clogged shallow coastal environments. Carbon-rich biomass sank to the bottom to be buried, even as the oxygen content of the atmosphere and oceans gradually rose.

Earth reached a tipping point somewhat more than 2 billion years ago: the "Great Oxygenation Event," when the accelerating intertwined cycles of carbon and oxygen set Earth on a path to the modern world. Dependent as we are on the oxygen-rich atmosphere, it's easy to forget that the production of oxygen by the modern two-stage version of photosynthesis is a mere by-product—a form of chemical waste that played no further role in life's evolving story for a very long time.

By contrast, the Great Oxygenation Event did play a remarkable role in the evolution of the geosphere. By 2 billion years ago, the atmosphere had become richer in oxygen, perhaps 1 or 2 percent of modern levels. Many rocks began to react with the corrosive gas to produce a flood of new mineral species never before seen on Earth (or anywhere else in our solar system, for that matter).[36]

Prior to the Great Oxygenation Event, most minerals featured metal atoms in their more reduced, electron-rich states. Relatively common elements like iron and manganese, along with a host of rarer metals, such as copper, nickel, molybdenum, and uranium, were concentrated in no more than a few hundred mineral species. The flood of oxygen pouring into the atmosphere changed that scenario: thousands of new types of minerals appeared, as oxygen gobbled up every spare electron it could find.

My colleagues and I have estimated that two of every three mineral species on Earth, including many of the most beloved colorful crystals displayed in the world's natural history museums, are direct consequences of oxygenic photosynthesis. Most copper minerals, including deep-green malachite, rich-blue azurite, and semiprecious turquoise, appeared only after the Great Oxygenation Event. More than 90 percent of the almost 300 different kinds of uranium minerals, many of them in brilliant yellow and orange crystals, are also an indirect consequence of photosynthesis. Carbon-bearing minerals also exploded in diversity in response to Earth's newly evolved near-surface chemical environment.

The realization that Earth's mineral evolution depends so directly on biological evolution is somewhat shocking. It represents a fundamental shift from the viewpoint a few decades ago, when my mineralogy PhD adviser told me, "Don't take a biology course. You'll never use it!"

Variation 3: Big Life Emerges[37]

Close your eyes and imagine "life."

I'd be willing to bet that you thought of something big—maybe your cat, or a flower, or the wren at your bird feeder. A life-form even larger may have come to mind—a vision of a favorite tree, a panda, or perhaps an elephant. From the menagerie of Noah's ark to modern urban zoos, charismatic megafauna receive the most press. In the context of Earth history, that's a highly distorted view of life.

For at least the first three-quarters of Earth's history, "life" consisted exclusively of microscopic cells mostly hidden deep beneath the surface, only occasionally revealed as stony stromatolite mounds or smelly masses of green algal filaments. You would have needed a powerful microscope to gain any understanding of that ancient living world. Today's biosphere, filled as it is with swimming, crawling, and flying things, is a relatively modern innovation, representative of only the past 10 percent or so of Earth's rich evolutionary history. And that fact raises a question: Why, after 3 billion years of successfully living the single-celled lifestyle, would cells begin to cooperate in such a way as to make these large organisms possible?

The simplest answer is that it's difficult for any single cell to do it all—to make or eat every essential molecule of life, to protect itself from other hungry cells, and to make exact copies generation after generation. That's why nature's most primitive single-celled organisms often live in complex communities called "consortia," in which different kinds of cells take on specialized chemical roles. Microbial consortia perform an exquisite electron dance, always passing electrons between a donor and an acceptor. Some microbes get their energy from the Sun; others, from the chemicals produced by their photo-synthetic neighbors. Many consortium members develop specialized chemical skills, producing only a handful of essential biomolecules for the other members of the club. Consequently, cells in consortia become utterly dependent on their neighbors for survival.

These clever clusters of cells are not unlike our energy-driven economy. Some hearty souls produce energy by digging coal, growing food, gathering sunlight, or harnessing the wind. Others specialize by making useful products—cars, clothes, houses, music—and exchange those products for energy. Likewise, cellular consortia consist of many cells, each an independent contractor playing its own role, which is based on its distinctive genetic identity and internal chemistry.

Playing Well Together

A strikingly new and influential variation on this idea of cellular coop-eration emerged at least 1.5 billion years ago, when a group of rel-atively large single-celled organisms called "eukaryotes" (Greek for "true nucleus") began to feature internal structures bounded by their own membranes.[38] These "organelles" (somewhat analogous to vital organs in people) include the nucleus that houses the cell's DNA, mitochondria that act as the cell's power plants, and light-harvesting chloroplasts that convert light into energy-rich sugar. Some biologists view the rise of eukaryotes as the single most important innovation in life's history, for it provided cells with an internal energy source that allowed them to diversify and cooperate as never before.

How did this new complex cellular architecture of eukaryotes emerge? Mitochondria and chloroplasts provide critical clues. They have their own membranes and their own DNA, and they undergo their own replication, just as if they were independent cells living inside the larger eukaryote. The consensus today is that eukaryotic cells arose when a larger cell swallowed one or more smaller cells. Rather than the guests being digested by the larger organism, new cooperative arrangements were found.

That conceptual breakthrough, most famously championed by the brilliant and controversial biologist Lynn Margulis, is now conventional wisdom, retold with elegant illustrations in every introductory biology textbook.[39] It wasn't always so. For two decades, the biological establishment reviled the idea of "symbiogenesis." Margulis's 1967 paper advocating the concept was rejected more than a dozen times; her grant proposals were also rejected, with scathing reviews.[40]

The vehemence of these early criticisms stemmed in part from perceived threats to established evolutionary theory. Darwin's paradigm of evolution by natural selection required gradual changes owing to countless, usually minor mutations, followed by selective winnowing among diverse populations of more and less fit individuals. In the twentieth-century version of Darwinism, those mutations arose exclusively from genetic variations in DNA. In stark contrast, symbiogenesis posited that new life-forms sometimes arise by the cooperative merger of two completely different species. For a time there was a bitter impasse, resolved only by the discovery of DNA in mitochondria and chloroplasts and the clear implication that those organelles had once been independent cells.

Yet even as the bioscience community came to embrace the symbiotic origins of eukaryotes, rewarding Margulis with dozens of prizes, medals, and honorary degrees, she pushed the envelope. Symbiosis, she argued, is *the* principal driving force of life's evolution. Genetic variations result primarily from the transfer of DNA among cells, not mutations. While promoting her views, she attacked the neo-Darwinists, who, in her words, "wallow in their zoological, capitalistic, competitive, cost-benefit interpretation of Darwin."[41]

Margulis saw symbiotic evolutionary pathways everywhere—in termites, in cows, in trees, and in people. Virtually every cell on Earth, she would argue with deep passion and reams of anatomical and molecular data, depends critically on some other group of cooperating, not competing, cells. Without specialized organs packed with cellulose-eating microbes, termites and cows would die. Without their expansive symbiotic networks of root fungi, not to mention the diverse menagerie of soil microbes, trees would die. Without our rich gut biome of microbial companions, we, too, would die. Indeed, Margulis framed all of nature in this new symbiotic context: the Gaia hypothesis that Earth itself functions as a single self-regulating system.

In 2011, just a few weeks before her untimely death by stroke, I met with Lynn Margulis at her home in Amherst, where she was a professor of geosciences at the University of Massachusetts. Margulis was a force of nature: aggressively inquisitive, always questioning convention and creatively reframing nature. There was an unfiltered joy in her exploration, and it was clear to me that she just wanted to know how nature works and refused to accept conventional explanations of complex problems.

Margulis lived adjacent to the historic Amherst home of Emily Dickinson; as we walked and talked, she would interrupt herself, reciting verses of Dickinson's spare poetry, pointing to the exact narrow lane or row of shrubs referenced in the lines. Margulis had invited me to Amherst in part to discuss her growing conviction that evolution by symbiosis could be extended to previously unconsidered aspects of Earth's geosphere. The conversation in some ways anticipated the core ideas of mineral evolution, though I didn't realize it at the time. If only I could have shared those ideas with her.

Enigma Variations[42]

For the better part of a billion years—most of Earth's Proterozoic Eon, which spanned the vast interval from 2.5 to about 0.5 billion years ago—eukaryotic cells lived solitary lives. Sandwiched between the vibrant biosphere of today and that robust microbial realm of a billion-plus years ago was a brief, enigmatic time of emerging novelty—the Ediacaran Period, when the first complex multicellular organisms appeared. Soft-bodied animals have a much smaller chance of preservation as fossils in the rock record compared to their hard-shelled counterparts. Nevertheless, worm and jellyfish fossils are found world-wide in fine-grained, oxygen-poor rocks, where dead individuals are more likely to be mummified.

The earliest odd examples of big soft-bodied fossils that you can hold in your hand appear in rocks dating from about 575 million years ago—almost 35 million years before the widespread appearance of animals with mineralized shells. Ranging in size from small coins to dinner plates, these odd, rounded and frond-shaped creatures must have lived in garden-like settings on the ancient ocean floor. Debates regarding their identities abound. Most researchers think they were animals, perhaps something akin to sponges or jellyfish, though it's just as likely that they have no close living relatives to reveal their affinities. Others have posited that they are early photosynthetic plants or even a primitive form of lichen. The ongoing arguments add to the allure of the Ediacaran mystery.

Well-preserved examples of Ediacaran fossils are few and far between, often located in remote and forbidding places, including the scorching outcrops of Death Valley in California, the bear-infested wilds of northwest Canada's Mackenzie Range, remote cliffs in Arctic Norway, and politically tricky sites in Iran and Siberia. Given the easily accessible abundance of more recent fossil shells or the seductive lure of dinosaur bones, it takes a special breed of paleontologist to zero in on the scrappy remains from the Ediacaran Period.

Michael Meyer, a recent colleague at the Carnegie Institution and now at the University of Harrisburg in Pennsylvania, has taken on the challenge.[43] First impressions of Meyer are not of an adventurous world traveler. He sports wild Hawaiian shirts, decorates his office with *Star Wars* images, and uses cute photos of his infant daughter Sam as his screen saver. You could be forgiven if you took Meyer for a regular nine-to-five kind of guy. But then you find him working in his office late into the night, computer monitor crowded with tables and graphs, surrounded by sample boxes labeled "South Africa" and "Huangling, China." He casually remarks about having had "an inordinate number of potentially fatal animal encounters . . . manatees, alligators, sharks, lions."

Following PhD studies at the University of South Florida, and field studies in southern China, Argentina, and the Flinders Ranges of Australia, Meyer came to my predator-free lab to study ancient life. Our ambition was to create new fossil databases and expand old ones. Paleontologists were way ahead in this game, having spent decades building the Paleobiology Database, a repository with hundreds of thousands of records on fossils from around the world.[44] Nevertheless, the database wasn't complete, especially for specimens from before the "Cambrian explosion"—the time about 540 million years ago when shells became common. Working closely with colleagues Drew Muscente and Andy Knoll at Harvard University, Meyer was tasked with expanding the Paleobiology Database to include all Ediacaran fossils.

A year's labor of scouring the published literature and contacting experts from around the globe led to a catalog of 95 fossil types from almost 100 localities worldwide.[45] Armed with such a comprehensive tabulation, Meyer and Muscente were able to see previously hidden trends. Experts divide the Ediacaran Period into three stages. The earliest, Avalonian, stage began about 575 million years ago, shortly after the "Gaskiers glaciation," a time when some experts suggest that most of Earth's surface from poles to the equator was covered in a layer of ice and snow. The 15-million-year interval after the global Ice Age saw a significant rise in global temperatures and the appearance of the first frond-like creatures in sediments that suggest a relatively deep ocean environment.

The subsequent animals of the White Sea stage, a 10-million-year interval lasting from 560 to 550 million years ago, contrasted with the Avalonian fauna. White Sea fossils display a dramatic increase in diversity, with dozens of genera sporting fronds and branching structures living alongside a variety of flattened animals that look something like elaborately grooved pancakes or miniature air mattresses.

The third and final Ediacaran interval, the Nama stage, from 550 to 541 million years ago, saw a shift to near-shore sediments and a range of creatures, including more than a dozen tubelike forms and strange, elongated and folded shapes reminiscent of a primordial taco.

Armed with new data, Meyer and Muscente were poised to make a striking discovery. Graphing all of their data into a weblike network, with dots for species and lines connecting species that lived together, they illustrated the entire Ediacaran fauna in a single, revealing "network diagram." The picture that emerged was of two distinct clumps of organisms, with all of the Avalonian creatures clustered at one end of the network and the mingled White Sea and Nama faunas in a larger central cluster. Only three genera of Ediacaran fossils were shared by both clusters.

What Meyer and Muscente saw was stark evidence for a massive faunal turnover—a time 560 million years ago when most of the Avalonian animals disappeared and new organisms took their places.[46] The jury is still out on how dramatic and sudden this turnover might have been. On the one hand, fossil-bearing rocks of Avalonian age tend to reflect an offshore environment on the continental shelf, in contrast to the shallower, wave-buffeted sediments of White Sea and Nama fossils. So the faunal turnover might just reflect a relatively benign shift in zip code, from deeper to shallower water. On the other hand, the sharp contrast between the older Avalonian and younger White Sea faunas could indicate a more dramatic mass extinction event—what would be the earliest such global loss of life preserved in the fossil record. (Even Meyer and Muscente can't quite agree on that point.) What the Ediacaran story shows beyond doubt is that we have a lot to learn about life's 4-billion-year evolution.

The earliest eukaryotes represented a signal advance in the complexity of cellular life—bigger, more diverse, and with a greater chemical repertoire than any previous life-form—but they were still single-celled organisms. By contrast, in a multicellular worm or jellyfish or frond-like creature, different kinds of cells have to cooperate and specialize in very specific ways. Some cells are on the inside, while others decorate the outside. Some cells form the tips of fronds, while others cement themselves to the ocean bottom. Look more closely and you'll find that cells play different chemical roles as well—gathering food, digesting nutrients, distributing essential biomolecules, and expelling waste.

A fundamental rule of the game of life is that innovations must be beneficial. So why, after 2.5 billion years of seemingly stable single-celled life, did consortia of cells begin to stick together and play such specialized roles? Such life-forms face at least three challenges not experienced by single cells. First, their cells have to stick to each other in an orderly, structured way; most multicellular organisms need a head and tail, or a top and bottom. Then those cells have to cooperate in their uses of atoms and energy. The specialized cells that gather food must share their bounty with everyone else. And finally, like all life-forms, the cells of this community have to figure out a way to make exact copies of themselves. Those are significant hurdles—challenges that wouldn't have emerged unless some advantage was to be gained.

Another challenge of multicellularity is energy. A concentrated mass of cells, especially one with cells that play specialized functions, requires a relatively concentrated form of energy. Every cell is a tiny electrical circuit that needs a flow of electrical charge. That's why almost all multicellular life-forms on Earth depend on the concentrated chemical energy of oxygen. By comparison, hydrogen or sulfur can't provide a sufficient energetic boost for multicellular life. Every cell in an animal requires a stable oxygen supply, so cells on the insides might appear to be at a distinct disadvantage. At least two coping strategies emerged. In some primitive animals, the cells form folded layers with passageways that allow environmental oxygen to reach every cell. In effect, every cell is on the outside. More advanced animals, like us, rely on intricate circulatory systems, with blood as the highly specialized oxygen delivery system.

In spite of these challenges, multicellular organisms evolved and radiated with remarkable rapidity. New strategies for survival emerged, driven by increasing competition for resources—food, territory, and protection—as animals learned to eat plants and other animals. In the process, living cells played ever-more-active roles in Earth's dynamic carbon cycle.

Variation 4: Life Learns to Make Minerals[47]

Life has always been a story of survival: find food, make babies, avoid being eaten. For the past half-billion years, the biosphere has experienced an escalation of this competitive evolutionary saga—a literal arms race of pointy weapons and protective armor. It all began when cells learned to make minerals.

No one knows quite when or where the first shell arose. The roots are ancient. More than 3.5 billion years ago, single-celled colonies built "stromatolites"—those odd, domed mounds of carbonate minerals. Crude, mineralized plates and haphazardly armored burrows appeared soon after the earliest signs of multicellular life, perhaps 600 million years ago. But the explosion of life with elegantly sculpted hard parts—spiral shells, branching corals, serrated teeth, and intricately patterned bones—dates back to a narrow window in the early Cambrian Period, about 540 million years ago. Carbonate minerals came first, though details of this amazing mineralizing trick remain enigmatic. First, cells had to create a local chemical environment in which calcium and carbonate ions in solution combine to form a tough, encasing crystal. Then the mineral ingredients had to be aligned just so to make a functional protective home. How did it happen?

Understanding this biochemical innovation is the life's work of Patricia Dove, University Distinguished Professor in the Department of Geosciences at Virginia Tech in Blacksburg, Virginia.[48] Dove has fun with science. She'll hand you the delicately sculpted shell of a chambered nautilus or pull an egg out of her pocket to illustrate the everyday marvel of biomineralization. She has never lost the sense of wonder that was instilled in her as a girl growing up on her family farm in Bedford, Virginia, about an hour's drive east of Blacksburg. It's a familiar story for many successful scientists: parents and teachers who were encouraging, a love of nature and of building collections, and winning prizes at science fairs and scholarships to college—in Dove's case, Virginia Tech. A PhD from Princeton followed, with stints at Stanford and Georgia Tech, before she returned to her beloved Virginia.

In all of her studies, Patricia Dove emphasizes that shells, teeth, and bones are much more than simple mineral crystals. They always incorporate layers and fibers of proteins and other biomolecules that add strength and flexibility—features that inspired the design of light-weight fiberglass and carbon fiber composite materials. In some shells, the resulting mineral-protein composite is a thousand times stronger than the pure mineral. Dove also reminds you that biominerals play many roles beyond shells, bones, and teeth; they can serve organisms as lenses, filters, sensors, and even tiny internal compasses.

Dove's dynamic research team focuses much of its energy on the atomic-scale mechanisms of carbonate formation—a molecular dance based on intimate interplays between organic and inorganic carbon chemistry. The team has discovered that biominerals form when cells create specialized compartments—local environments in which the mineral-forming ingredients are concentrated, nucleated, and grown with exquisite precision. Within those compartments, some biomolecules promote crystal formation, while others inhibit growth.

One of the most surprising discoveries by Dove's research team is that many organisms initiate biomineralization with a form of calcium carbonate that is not crystalline. Instead, they form and store a gel-like substance, amorphous calcium carbonate (they call it "ACC") that is held in reserve until just the right moment.[49] A molecular trigger starts crystal growth in this remarkable process, in sharp contrast to conventional crystal formation. Some molting animals can apparently store ACC for periods of weeks or months, triggering rapid shell growth at the critical vulnerable stage when an old shell is discarded and tasty soft tissue is exposed.

For a time after the first precocious animals armored themselves, predators must have turned most of their attention to easier, unprotected morsels. Why expend all the extra effort to crack a tough shell when fleshy worms were close at hand? But as more of the seafloor denizens became armored, it didn't take long for counterstrategies to emerge; stronger jaws, sharper teeth, and more vicious claws all appeared as life learned to build protective shells. This protracted Cambrian "explosion" wasn't all that explosive—it lasted tens of millions of years—but it set Earth's biosphere on an irreversible course.

The invention of tough mineral shells also added a new wrinkle to the carbon cycle. With the rise of carbonate corals, bryozoans, brachiopods, mollusks, and other fauna, limestone reefs achieved epic proportions, spanning hundreds of miles of coastlines and, ever so gradually, in some places reaching thicknesses of thousands of feet. Never before had such massive accumulations of carbonate biominerals been possible, filling shallow coastal waters and inland seas with unprecedented deposits. When plate tectonics inevitably closed those shallow bodies of water and compressed their sediments, jagged mountain ranges capped by carbonates altered Earth's landscape. The Canadian Rockies, the Dolomites of northern Italy, even the highest prominences of Mount Everest and other Himalayan giants are made of sturdy carbonate— living reefs that once graced shallow ocean floors.

—

For most of the past half-billion years, carbonate biomineralization was almost exclusively a near-shore activity—the domain of the carbonate reef. Corals, snails, clams, and dozens of other life-forms take advantage of shallow, sunlit, and nutrient-rich waters along the coasts of continents.

Two hundred million years ago, life discovered yet another mineralizing trick. Microscopic single-celled marine algae called coccolithophores, which today flourish in all of the world's oceans often far from any landmass, learned to make tiny, transparent, disklike armored plates of calcium carbonate called coccoliths.[50] Each coccolith is like a microscopic ornamental hubcap less than a thousandth of an inch in diameter. For reasons that are not fully understood, a dozen or more of these overlapping plates cover each algal cell. Some biologists suggest that the mineral disks are armoring for protection; others, that the calcium carbonate is a natural sunblock that shields the floating cells from damaging ultraviolet radiation. Yet another hypothesis is that the mineral plates provide cells with neutral buoyancy, allowing the microorganisms to sink or float to more nutrient-rich layers of the ocean.

Whatever their function, coccoliths are produced in prodigious quantities. When coccolithophores die, their tiny plates accumulate in thick deposits of chalk. A microscopic view of Dover's famed White Cliffs reveals astronomical numbers of the beautiful sculpted forms— hundreds of feet of chalky sediments deposited over millions of years. Unlike the situation in previous eons, as much as a third of today's ocean floor is covered in calcareous ooze, in many places a mile or more deep, rich in microscopic disks.

The implications for Earth's carbon cycle are profound. For most of Earth's history, the sediments of the deep oceans were relatively free of carbonate minerals. Subduction recycled ocean crust that was dominated by basalt. Today, by contrast, carbon-bearing minerals are a major component of the seafloor. As the ocean floor is swallowed up in subduction zones, some of that newly minted carbonate ooze is carried along, plunging deep into Earth's mantle. An unsolved mystery is whether that burial of carbon has now fundamentally changed Earth's carbon cycle. If more carbon atoms are buried than return to the surface, could Earth's biosphere gradually become carbon starved?

Answering this profound question about Earth's deep carbon cycle leads us inexorably to additional feedback loops between the biosphere and geosphere, so we must shift our focus to the rise of life on land.

Variation 5: Life Establishes a Foothold on Land[51]

As life and rocks coevolved in ever-more-complex feedback loops, carbon cycling increased in complexity as well. In no sphere were these feedback loops more pronounced than in the rise of life on land.

Oxygen played a pivotal early role. The Sun's ultraviolet radiation, if unfiltered, is far too harsh for most cellular life to survive direct exposure. Key biomolecules fragment. Cells die. The rise of oxygen in the atmosphere also meant the rise of "ozone"—a molecule comprising three linked oxygen atoms that forms when ordinary O_2 molecules are split and rearranged by ultraviolet radiation. When that happens, some atoms recombine into O_3. Ozone molecules are sparse, representing at most a few molecules in a million when they "concentrate" to form the ozone layer in the upper atmosphere, about 20 miles up. Once that layer forms, it plays the role of a natural sunblock, shielding Earth's surface from the Sun's incessant UV blast. A healthy ozone layer is a prerequisite for a robust land-based ecosystem.

The earliest steps away from the sea were cautious, barely altering the landscape. First on the scene more than 450 million years ago were tiny, rootless plants that added a dash of green to the marshy perimeters of coastal pools and shallow streams. By 430 million years ago, the first plants with diminutive root systems had appeared, allowing green plants to establish new ecosystems farther inland. Roots accelerated the breakdown of rocks, forming clay-rich soils that supported longer and more efficient roots, which in turn produced more soil. In a geological flash, bushes and trees of increasing height and girth blanketed the land.

Change followed change. Land plants facilitated the radiation of animals—a primitive millipede, *Pneumodesmus newmani*, is the earliest known example of an air-breathing land dweller.[52] Found by Scottish bus driver and amateur fossil collector Mike Newman in 2004, the lone specimen is a half-inch-long fragment in 428-million-year-old sediments from Aberdeenshire, Scotland. The fossil record of such soft-bodied ancient animals is poorly preserved and exceedingly sparse, so it's likely that primitive insects appeared even earlier. The earliest centipede fossil dates from 420 million years ago, the oldest known flying insect from about 400 million years ago, and other rare and revealing treasures are surely awaiting discovery.

Vertebrate fossils of land animals, though more likely to survive in the rock record, are nevertheless few and far between. The modest (and growing) numbers of known species from the Paleozoic Era display a gradual transition from sea to land, from fish to amphibians, with increasingly specialized structures for life away from the seas. Fins morphed into toed feet with shoulders, elbows, and wrists. Skulls developed nostrils to breathe and ear openings to hear. And, unlike most fish, the earliest land dwellers had necks; they could swing their heads from side to side to survey their dry surroundings. The transition to land wasn't sudden, and it may never be possible to point to the "first" terrestrial vertebrate, but a strong contender for the title is the 375-million-year-old *Tiktaalik roseae*, discovered in 2004 on Ellesmere Island, north of the Arctic Circle in Canada's remote Nunavut province.

In a spectacular display of paleontological sleuthing, Neil Shubin of the University of Chicago and Ted Daeschler at the Academy of Natural Sciences in Philadelphia predicted that they might discover such a beast in northern Canada—a frigid Arctic location that, thanks to the perambulations of wandering tectonic plates, was close to the equator 400 million years ago.[53] Their prediction arose through a logical process of elimination. They realized that the "missing link" between fish and amphibians must occur in rocks about 375 million years old, preferably from a warm, equatorial region and near the ancient coastline. The locality also had to have good rock exposures; a deeply forested site would not do. They scoured geological maps of the world and zeroed in on Ellesmere Island as an ideal locality for further exploration.

Finding a missing link on a remote Arctic island wasn't an easy task. The locality is all but inaccessible, and collecting seasons are short—just a few weeks per year during midsummer, after the thick blanket of snow has melted and before the first new snows of the fall arrive. It took five fruitless field seasons, some focused on unproductive rock layers, others interrupted by terrible weather, before they discovered the first stunningly complete finds of the crossover fish/amphibian *Tiktaalik* in a low, rocky ledge.[54] It was a large creature, with some individuals growing to almost 10 feet in length. Seeing the entire beast, Shubin and Daeschler realized that fossils of this organism were rather common; several scrappy and unrecognized *Tiktaalik* fragments had been recovered in previous collecting seasons.

This walking fish, named after a native Inuktitut word for a local variety of cod (though informally dubbed the "fishapod" by discoverers Shubin and Daeschler), has become a media sensation, with public lectures, TV shows, and its own website. Shubin's popular account, titled *Your Inner Fish*, rose to the status of science bestseller with its own media spinoffs. The entire paleontological saga—from the bold prediction, to the difficult discovery and the growing realization that many of the anatomical innovations of *Tiktaalik* persist in structures in our own bodies—demonstrates once again the power of Darwin's theory of evolution by natural selection.

Tiktaalik was only one of a succession of fossil animals, each more adapted to full-time life on hard ground than the last. The pace of the transition was geologically rapid, though it would be another 10 million years before the first unambiguous land animals roamed the primitive jungles of Earth. All the while, as carbon was concentrated in roots, stems, leaves, and trunks, the cycling of carbon among varied reservoirs—Earth, Air, Fire, and Water—intensified.

Buried Biomass

Forests evolved to become the most exuberantly diverse new carbon reservoirs following the rise of life on land. They added a new wrinkle to the carbon cycle, as the giant plants of Earth's earliest swampy forests—vast swaths of luxuriant ferns, cycads, and conifers—pulled carbon from the air to make wood and bark. When one of those earliest land plants died, its stem, branches, leaves, and roots contributed biomass to form new types of carbon-rich sediments: loosely consolidated peat from near-surface bogs; soft, brown lignite; and the hard, black fossil fuel known as coal.[55]

Most of Earth's coal formed during a fleeting 60-million-year interval, starting about 360 million years ago—the aptly named Carboniferous Period. In today's forests, when a tree falls it usually rots away quickly, returning carbon atoms to the soil to be used again and again. Contrast that efficient recycling to 300 million years ago, a time before the evolution of the many varieties of "lignicolous" fungi, which mastered the trick of breaking down wood's tough lignin fibers. Before the advent of wood decay, dead trees piled up in layers 100 feet or more in thickness. Plant remains were buried deeper and deeper, their tissues compressed and baked. That biomass gradually dried out, while biomolecules depolymerized, releasing volatiles and increasing the carbon content to more than 90 percent in the most desirable varieties of anthracite coal. Today we mine that Carboniferous legacy at prodigious rates, returning 60 million years of sequestered carbon to the atmosphere in a matter of decades.

Even as layers of coal accumulated, Earth's growing landscape of rich, deep soils found another way to sequester carbon. Clay minerals, the abundant by-products of probing roots and their inexorable destruction of rocks to soil, began to play a significant role as well.[56] Clays are unique in their physical and chemical behavior. Clay minerals form as thin, flat mineral plates too small to see in an ordinary microscope. These minuscule sheets slide over each other, giving clays their familiar slippery-when-wet character.

Clay surfaces are also exceptional in their ability to bind to small, carbon-rich molecules, including the decay products of life. When roots and other subsurface detritus rot, their biomolecules are, as often as not, sequestered on clay mineral surfaces. Soils erode as rivers and winds transport huge quantities of clays to the oceans. Thousands of feet of sediments accumulate offshore, and those sediments hold a lot of carbon—yet another reservoir in Earth's intricate carbon cycle. And some of those carbon-rich sediments subduct deep into Earth's mantle along with countless carbonate coccoliths—fluxes of carbon that may set the age of life on land apart from the previous 4 billion years of Earth history.

Variation 6: We Make Our Mark

Countless millions of species have arisen, and the vast majority of those life-forms have disappeared forever. Trilobites, the ubiquitous, charismatic denizens of Paleozoic seas, emerged more than a half-billion years ago. Their prized fossilized remains, segmented and spiny, seem to stare at us with dimpled, compound eyes from across the ages. All have died—extinct for more than 250 million years. Dinosaurs on land, at sea, and in the air had their turn, ruling the Mesozoic world with grandeur and savagery. Their massive skeletal remains provide silent reminders of nature's relentless struggle for survival. All have died, save for the birds—a drastic winnowing of a once dominant lineage. Now it's our turn.

The human story is more sharply etched than those of other species. We humans change the environment in ways that last. We build monuments, we dig coal, we light fires, and we leave objects behind. In that ancient story, carbon plays a special, surprising role, for even as we live our lives and build our culture, carbon atoms provide us with a clock to chronicle the human story.

The Carbon Clock

Almost every carbon atom is the lasting legacy of stars—permanent, unchangeable, to be used again and again. But a tiny, tiny fraction of the carbon atoms in the air and in our bodies are transient players on Earth's dynamic stage. These atoms appear as if by magic and strut their stuff for a short while, only to vanish in a whiff.

We've already met carbon's two stable forms: common carbon-12 comprises more than 99 percent of your body's carbon atoms; its slightly overweight cousin, carbon-13, accounts for the remaining 1 percent. These isotopes—one with 6 neutrons, the other with 7— formed billions of years ago, primarily in big stars.

Radioactive carbon-14, with 8 neutrons, is different.[57] It is unstable and ephemeral. Carbon-14 forms continuously high above most clouds in a realm where cosmic rays from deep space bombard the nitrogen-rich atmosphere. Cosmic rays, most of them speeding protons or atomic nuclei, act as energetic atomic bullets that collide with atmospheric molecules, causing nuclear chaos. Showers of secondary particles spray outward, some of them energetic neutrons, some of which smash into nitrogen atoms. When struck by a fast-moving neutron, a nitrogen-14 nucleus can be disrupted, losing one proton while gaining one neutron to form carbon-14. This violent, creative process, ongoing for billions of years, produces a small but continuous supply of carbon-14 in Earth's atmosphere.

The critical difference between carbon-14 and its lighter, more stable carbon cousins is radioactivity. Carbon-14 teeters on the brink of self-destruction because it has too many neutrons for comfort. Without warning, a radioactive carbon-14 atom will spontaneously change back into a stable nitrogen-14 atom. Carbon-14's radioactive decay is a reliably gradual process, taking about 5,730 years for half of any collection of radioactive carbon atoms to vanish. This fortuitous "half-life" is ideally suited for the study of evolving human technology and culture by the powerful method of "radiocarbon dating."

The transformative technology of carbon-14 dating depends on death, or more accurately, the time of death. Carbon cycling is key. As long as a plant is alive, it constantly takes in carbon dioxide, water, and the Sun's radiant energy to make sugar. That's the essence of photosynthesis, which provides chemical energy for almost all life on Earth. Animals eat the sugar-rich plants, or they eat other animals that eat the plants. Fungi and scavengers consume dead plants and animals. At every step in the complex food web, carbon atoms cycle from one reservoir to the next.

As long as a plant lives, it takes in about 1 part per trillion carbon-14 along with other carbon atoms—the ratios of carbon-12, -13, and -14 being fixed primarily by the atmosphere. As long as you eat plants, or eat animals that ate those plants, you, too, will share that isotope ratio: about one in every trillion carbon atoms in your body is radioactive carbon-14. And that tiny fraction will remain more or less constant until the plant dies, or until you die. That's when the carbon clock starts ticking.

Storytelling

The idea of using radioactive carbon to date the remains of life was the inspiration of University of Chicago chemist Willard Libby shortly after World War II.[58] As a scientist with the Manhattan Project, Libby was well versed in the chemical behavior of radioactive isotopes, and he realized that carbon-14 held special promise for investigating the recent history of human civilization. Like so many of his atomic-era colleagues, he turned his expertise to nonmilitary applications after the global conflict.

Libby's concept is simple: Take an old parchment, a piece of charcoal from a wood fire, a fiber of hair, or a flake of dried skin; measure the ratios of carbon isotopes; and calculate the age. If half of the carbon-14 atoms have decayed, then the object is about 5,730 years old. If only a quarter remain, then the age is double that—roughly 11,500 years old. Radiocarbon dating works remarkably well for fragments of past life as old as 50,000 years, after which only about a thousandth of the original radioactive carbon atoms survive.

In practice, carbon dating is a bit trickier. For one thing, accurately measuring 1 part per trillion of carbon-14 is no easy task. In one common approach, scientists count each radioactive decay event and calculate the carbon-14 content from the levels of radioactivity. Radiocarbon decay is slow, so that method requires big samples with lots of carbon atoms, as well as lots of patience. More efficient modern methods enlist powerful mass spectrometers to measure the amounts of heavy carbon-14 isotopes before they decay. It's a faster approach, and one that can be applied to much smaller samples, no larger than a millet seed or short strand of hair.

Radiocarbon dating has revolutionized our understanding of human history. You can see the consequences every week in the news.

The artifacts of Christianity have received special attention. A collection of dozens of scrolls written in ancient Hebrew and Aramaic, discovered in caves near the Dead Sea in 1947, provided a high-profile early test of Willard Libby's new dating technique. He showed them to be roughly 2,000 years old, and thus the earliest known biblical texts. On the other hand, the famed Shroud of Turin, revered by some believers as the linen burial cloth of Jesus of Nazareth, was shown in 1988 by three independent laboratories to date from the fourteenth century. The source of this beautiful object with its ghostly image of a man remains a matter of intense debate.

Radiocarbon dating also plays critical roles in archeology, providing detailed chronologies of Egyptian dynasties, the sequence of African migrations, the transfer of technology in Europe, and the settlement of prehistoric Britain. Carbon-14 reveals the ages of innumerable prehistoric sites and objects, from monolithic Stonehenge, the earliest portions of which are dated at 5,100 years by buried pieces of wood, to Ötzi the "iceman," who died 5,200 years ago and was preserved in alpine ice near the border between Austria and Italy. The discovery and dating of artifacts is also changing the Eurocentric view of the "discovery" and colonization of the Americas, with clear evidence from campfires of Viking settlements as early as 1000 AD, five centuries before the first voyage of Columbus.

Radiocarbon dating also plays a central role in constraining the controversial chronology of human migration to the Americas. In a 2015 study published in the *Proceedings of the National Academy of Sciences*, scientists from Texas A&M University described an ancient campsite with the bones of butchered horses and camels discovered near Calgary in Alberta, Canada.[59] At a carbon age of 13,300 years, with an uncertainty of only about 15 years, the site is older than those documented for the Clovis people, who are thought to have arrived from Russia across the Bering Strait no earlier than 13,000 years ago. Other, less certain evidence in the form of campfires associated with primitive stone artifacts leads some researchers to postulate even older ages for a migration from Asia to North America—perhaps as far back as 40,000 years. Whatever the eventual conclusion, radiocarbon dating will play a key role.

—

What of the coming generations? What insights regarding modern times will future archeologists glean from the carbon residues of our own age? They will find surprises. A significant influx of "dead" carbon distinguishes the past two centuries—the legacy of burning vast quantities of fossil fuels that contain ancient carbon atoms locked away for millions of years. The resulting flood of dead carbon dioxide dilutes the atmosphere with molecules of CO_2 completely lacking radioactive carbon-14.

A second, even more dramatic "bomb carbon" anomaly marks the brief frenzied era of open-air nuclear weapons testing—the time in the 1950s and early 1960s before the Nuclear Test Ban Treaty went into effect.[60] In little more than a decade, nuclear explosions caused the atmospheric concentration of carbon-14 to double, only to gradually decline as carbon dioxide in the altered air traded places with molecules in the oceans, became sequestered in rocks, or were consumed by plants. Inevitably, for a short time, the carbon-14 content of plants doubled, followed by a doubling in animals, and in you if you were alive during that troubled Cold War era.

All of us still share a bit of that nuclear legacy in our muscles and in our bones, for we are all part of the human carbon cycle.

RECAPITULATION—The Human Carbon Cycle

THE KEY TO UNDERSTANDING the inextricable connections among Earth, carbon, and us is cycles. Carbon cycling lies at the heart of the rapid changes that humans are imposing on Earth, both planned and unintentional. We grow and breed carbon-rich foods to support an expanding world population, inevitably disrupting environments that have been stable for thousands of years. We strip the land of forests and the seas of fish, altering the ecological balance. We mine carbon in vast quantities, exploiting long-buried resources for fuel and material goods. In each of these examples and more, the accelerating human impacts on the carbon cycle are global and profound, and the consequent changes in the atmosphere and climate are largely unintended.

All living things, our human species included, play a role in the global carbon cycle. Look around you and the consequences cry out: Earth's minerals and air become plants; plants become the food for animals; dead animals and plants support thriving fungi and microbes; in turn, all these organisms return to the soil and the mineral realm. Carbon atoms cycle again and again; each atom experiences many forms in its billions of years of existence.

—

You and I also experience a personal carbon cycle—the much more immediate and intimate cycle associated with our changing bodies. From the moment of embryonic conception to the time when our dead remains decay away to nothing, we experience this personal cycling of Element 6.

You inhale oxygen and eat carbon-rich food, which are the reactive drivers of metabolism. Your body incorporates that carbon as it builds new cells; your body generates carbon dioxide as it burns carbon-rich fuel.

You exhale carbon dioxide—your body shedding carbon atoms like tree leaves in autumn. With each breath you dissolve just a bit.[61] With each breath a tiny fraction of your body's carbon—less than a thousandth of a percent—is lost, dispersed, poised to recycle. Your body today may seem the same as your body of last week or last year, but it's not. Many of the atoms are different—exact atomic copies, but different.

All of your life you have consumed new carbon atoms while shedding old carbon atoms. How ephemeral our bodies are! Few of the atoms, the molecules that were you at birth, remain. Few of the atoms, the molecules that are you today, will still be you if you should live another decade. We humans are quick to equate our bodies with the essence of ourselves. Our minds are separate, our thoughts uniquely our own, but the atoms in our bodies are as fleeting as the breeze.

And where are those atoms now—the atoms that were, until a short time ago, part of you? Some are in the air or dissolved in the oceans. Some may be locked in the carbonate shells of clams and snails, or soon to be sequestered in the limestone of coral reefs. Many of the trillions and trillions of carbon atoms that were once you now reside in the stems, leaves, flowers, and roots of plants—oak, wheat, rose, moss. Animals eat the plants and so inherit and hold for a brief time what was once you. And every person on Earth who lived more than a few years, anyone who has eaten the plants, or the animals that ate the plants, now holds carbon atoms that were once you, even as you hold carbon atoms that were once them—atoms of everyone you have ever known—your friends, your family, your lovers, almost everyone who has ever lived.

—

Imagine the cosmic path of a single carbon atom that is now, for a fleeting time, part of what you think of as "you."

That carbon atom was forged in the heart of a big star, released to space when that star exploded. Let's say it joined with other carbon atoms to form a tiny crystallite of diamond, to become part of the dust and gas in a molecular cloud—a rich, star-forming region of our young Milky Way galaxy. A perturbation in the cloud, perhaps a jolt by a nearby supernova, triggers a local collapse, the beginnings of our solar system. Most of the mass falls inward to make our Sun, but the diamond speck finds a different home in the leftovers that begin to form the third big planet in the consolidating system.

Earth is too hot, too reactive for the little diamond to survive. The carbon atom links with oxygen atoms to form a molecule of carbon dioxide—a minuscule part of the growing atmosphere. The CO_2 molecule absorbs into the ocean, where it follows the currents for a thousand years before precipitating as a carbonate coating on the shallow bottom of an ocean margin.

Another million years, 2 million, pass. Then total disruption as a mile-wide asteroid blasts the coastline. Carbonate minerals vaporize, returning carbon dioxide to the atmosphere. The cycle repeats—air to ocean, ocean to rock—but this time the layer of carbonate mineral is trapped in a descending fragment of dense crust, slowly plunging into the upper mantle, where Earth's inner heat melts the surrounding rock. That melt, rich in water and CO_2, rises ever closer toward the surface, pressure confining the volatile mix. As the magma approaches the surface, the fluids turn suddenly and violently into explosive steam, blasting outward, showering the landscape with volcanic boulders and ash. Once again the carbon atom is liberated into the air as a molecule of CO_2.

Our well-traveled carbon atom finds itself energized by a nearby bolt of lightning; it combines with nitrogen and other atoms to form an amino acid—a molecule that survives only a few days before being fragmented by the Sun's ultraviolet radiation. In its more stable guise as part of carbon dioxide, the atom cycles repeatedly from air to ocean. More than once, the atom undergoes reactions at a deep ocean vent, forming amino acids that survive only a few weeks before decomposing back to CO_2. And so the star-born carbon atom cycles through Earth's reservoirs for eon after eon, from gas to liquid to rock and back again, in time without number.

Fast-forward a billion years. The new phenomenon of life emerges. New carbon reservoirs beckon. Carbon dioxide is pulled from the air and converted to sugar by photosynthetic algae. Sugar becomes the fuel for the manufacture of remarkable molecular novelties: carbon-chained lipids to form the cell membranes; carbon-ringed bases to carry the genetic code; carbon bonded to nitrogen and oxygen in amino acids, the building blocks of proteins. Our carbon atom cycles quickly through the biosphere, playing many new roles at a pace far more frenzied than before—sometimes changing chemical form a dozen times in a week. Other times it is sealed into a carbonate shell, sinking to the ocean floor, sequestered for 100 million years before returning to the vibrant, living surface world.

Last week you ate that carbon atom. It is now part of a protein molecule that serves a vital role in one of your cells. Let's hope nothing goes wrong.

Death and Carbon

We multicellular life-forms are vulnerable. A lot can go wrong with carbon atoms in such complex systems.

My brother Dan's cancer began in his duodenum, a place no one looks for cancer, a place where no acute symptoms arise, at least not until the cancer has spread to other organs. Then it's too late. Doctors tried to arrest the aggressive cells, as the cancer killed his liver. Dan endured months of ghastly chemo. No use. Dead in half a year.

Carbon-based molecules were the culprits, as they are in many diseases. A carbon atom is a little thing, but a few carbon atoms missing, misplaced, or misaligned can make a great deal of difference: we are subject to the whims of carbon in death as surely as in life. Doctors couldn't tell us what specifically went wrong with Dan—why he, the fittest of us all, most careful of diet, most dedicated to exercise, should succumb. Something went wrong with the molecules that control cell division in just one of his tens of trillions of cells. That cell began to reproduce out of control, ultimately to usurp other cells, other organs.

All cancers and genetic diseases are the same in that respect: mistakes in carbon atoms—their locations, their bonds. Consider the essential amino acids aspartate and glutamate, which differ by just one carbon atom. Insert the wrong amino acid into a critical protein and the long chain of molecules folds one way when it should have folded another way. The resulting malformed structure can wreak cellular havoc, causing the cell to fail in its life-critical mission.

—

And when we die, where do the carbon atoms go?

I think of Lulu, our fourth and last Maltese, the third generation of sweet, white fluff balls. For most of her thirteen years she was a perky sprite, bounding out the door, barking and leaping when I came home from the lab. Her twin sister, Julia, had died a year earlier, aged twelve, and Lulu's decline was steady after that. Toward the end she was deaf, confused, alternately hunkered down and racing jerkily about, barking at phantoms. When she stopped eating and drinking, we had her put down. At least the end was peaceful, the once lively home overwhelmed by silence.

We dug a grave in the woods near the house, beneath a flowering redbud. Lulu's curly, pure-white hair seemed out of place in that deep, dark hole; we said goodbye as two feet of rich, brown earth covered her.

A small dog of 10 pounds incorporates 2 or 3 pounds of carbon, perhaps 50 trillion trillion carbon atoms. What happened to them

when Lulu died? For a short while—hours, not days—her dead body
held on to almost all of those atoms. But, exposed to air and soil, and
offering a rich store of chemical energy to bacteria, fungi, and small
scavenging animals, her atoms began the inexorable process of dispersal. Most of the dead flesh was consumed to provide energy and
atoms to other living creatures. Her carbon atoms began to diffuse
away, spreading outward in ever-greater circles of diluted Lulu. Decay
also released carbon dioxide and other small organic molecules to the
atmosphere, to be spread around the world, ultimately to be used and
reused innumerable times in new life on every continent. Even now
you may be breathing in atoms that were once part of Lulu.

Recycling. That's what nature does. Earth's stable carbon atoms
are neither created nor destroyed; they are used over and over again.

FINALE—Earth, Air, Fire, and Water

W HAT IS OUR ROLE in the evolutionary scheme of things, in the great carbon symphony? Humans are at once ordinary and unique. On the one hand, we are just another evolutionary step in a 4-billion-year story that will likely continue long after our lineage has gone extinct or morphed into some new species. Some argue that we alone have the ability to radically alter Earth's climate and environment, but oxygen-producing photosynthetic microbes and the diverse green plants that followed them have changed Earth's near-surface environment in ways far more profound than any human actions. Others point to humanity's global influence on the continents through building cities, roadways, mines, and farms, but trees and grasses far outstrip our impact on the landscape. Some say our species is unique in its potential to "destroy the planet," but repeated catastrophic impacts of asteroids and explosive eruptions of megavolcanoes have had far greater destructive consequences than any weapon devised by humans.

At the same time, our human species does possess unprecedented abilities. We are unique in the history of life in our technological prowess to adapt and alter our environments at scales from local to global. We are unique in our inventive exploitation of other species— animal, vegetable, and microbial. We are unique in our exuberant desire and ability to explore beyond our world, perhaps eventually to colonize other planets and moons. And we are unique in our impact on Earth's carbon cycle—a cycle that profoundly affects every aspect of our planet—Earth, Air, Fire, and Water.

Humans are unique among life-forms because of the frenetic pace of the changes we impose. We are altering the planet at rates much faster than any prior species has—at rates exceeded only by the sudden cataclysms of volcanoes exploding and rocks falling from the skies. Microbes took hundreds of millions of years to oxygenate the atmosphere, and perhaps a billion years more to oxygenate the oceans. Multicellular life required tens of millions of years to colonize the land after the earliest tentative encroachments. These changes were profound, but they occurred over geological timescales that enabled life and rocks to coevolve gradually. Earth's ecosystems are remarkably resilient, but they need generations to shift, to evolve, and to reset themselves in response to new environmental conditions. If humans pose a unique threat to Earth, as some scholars fear, then it is the unprecedented rate of environmental change that carries the greatest risk for damage to the biosphere.

That being said, the rocks and the varied microbes that live among them will do just fine, no matter what injuries we might do to our home and, inadvertently, to our own species. Earth will go on, life will go on, and the powerful process of evolution by natural selection will ensure that new creatures continue to inhabit every niche on the planet.

—

Carbon's grand, eternal symphony unifies all of the elemental essences—Earth, Air, Fire, Water. Nothing exists in isolation; all are essential parts of the whole. Earth grows the solid crystals of carbon—sturdy foundation stones of land and oceans alike. Air holds the molecules of carbon that embrace us all—forever cycling, protecting and sustaining life. Fire, born of carbon, energizes the world, while providing unrivaled molecular variety to the material and living worlds.

Water, which gave birth to carbon life, nurtures that life as it evolves and radiates to every corner of the globe. In a crescendo of exquisite harmony and complex counterpoint, each essence of carbon celebrates, and is celebrated by, the others.

Humans have learned to impose their own urgent themes and ever-accelerating tempi on this ancient score. We strip Earth of its minerals. We flood Air with our waste. We harness Fire to satisfy our wants and needs. We exploit the teeming, living sphere of Water, often careless about which species live or die.

We must, each of us, step back from the urgency of our desires to see our precious planetary home as a unique, but vulnerable, dwelling place. If we are wise, if we can temper our wants with a renewed sense of awe and wonder, if we can learn to cherish our rhapsodically beautiful carbon-rich world as it so urgently deserves, then we may hope to leave an unrivaled, priceless legacy for our children, their children, and all the generations to come.

Acknowledgments

N WRITING *SYMPHONY IN C* I have benefited enormously from my colleagues in the Deep Carbon Observatory. None of DCO's discoveries would have been possible without the initial contact and subsequent support and counsel of Jesse Ausubel and his colleagues at the Alfred P. Sloan Foundation, especially Program Officer Paula Olsiewski.

Rus Hemley, former director of the Geophysical Laboratory, was instrumental in defining the scope and content of the DCO, and he conceived the name "Deep Carbon Observatory." The effort was initially shaped by DCO director Connie Bertka, who imagined an organizational structure emerging with bottom-up enthusiasms structured by gentle top-down nudging.

I am deeply indebted to the staff of DCO's secretariat, with whom I've had the pleasure to work on a daily basis. Director Craig Schiffries has guided the complex international program with skill and humor, while surviving the sweltering, oil-rich plains of eastern China, the toxic exhalations of volcanoes in Italy, and the crumbling cliffs of Oman. Craig also provided constant advice and encouragement during the writing of this book, and his thoughts are reflected throughout. Program manager Andrea Mangum, the longest-serving member of the DCO team, has expertly handled DCO's financial health and has been a steadfast, calming voice. Jennifer Mays and Michelle Hoon-Starr have brought skills in writing, web design, and logistical support, coupled with inspiring enthusiasm and commitment to DCO success.

The DCO would not have succeeded without the dedicated international scientific leadership of the Executive Committee, led by Craig Manning of UCLA. I have learned from all of its members, including John Baross, Taras Bryndzia, David Cole, Isabelle Daniel, Donald Dingwell, Marie Edmonds, Peter Fox, Erik Hauri, Russell Hemley, Kai-Uwe Hinrichs, Claude Jaupart, Adrian Jones, Louise Kellogg, Karen Lloyd, Bernard Marty, Eiji Ohtani, Paula Olsiewski, Terry Plank, Robert Pockalny, Craig Schiffries, Barbara Sherwood-Lollar, Nikolay Sobelov, Mitch Sogin, and Vincenzo Stagno.

Numerous individuals provided their expertise in reviewing various versions of the manuscript. I am grateful to David Cole, Darlene Trew Crist, David Deamer, Patricia Dove, Marie Edmonds, Charlene Estrada, Paul Falkowski, Teresa Fornaro, Shaun Hardy, Grethe Hystad, Olivia Judson, Jie Li, Andrea Mangum, Scott Mangum, Craig Manning, Shuhei Ono, Sarah Rugheimer, Craig Schiffries, Eric Smith, Dimitri Sverjensky, and Edward Young.

I'm grateful to my DCO colleagues in our superb Engagement team—Rob Pockalny, Katie Pratt, and Darlene Trew Crist—who have helped immensely in the development of this project, and especially Joshua Wood, who played a critical role in the selection and design of the photo insert.

I owe a special debt to my editors and production team at W. W. Norton. Quynh Do edited the manuscript with thoughtful insight. She brought to the project a simultaneous appreciation for the sweep of the book and careful attention to detail. Copy editor Stephanie Hiebert enhanced every page with meticulous and creative improvements. I also thank project editor Amy Medeiros, production manager Julia Druskin, and art director Ingsu Liu.

My literary agent, Eric Lupfer of Fletcher and Company, has been a continuing source of sage advice and unflagging support. He was among the first to embrace the concept of a book as a symphony.

Science is expensive. None of the advances described in this book would have been possible without the support of government agencies and private foundations. I am indebted to the National Science Foundation, the US Geological Survey, and NASA, including the remarkable Mars Science Laboratory program and the NASA Astrobiology Institute (your tax dollars at work). For their generous support, I am grateful to not only the Alfred P. Sloan Foundation, but also the W. M. Keck Foundation, the John Templeton Foundation, the Simons Foundation, the Gordon and Betty Moore Foundation, and the Carnegie Institution for Science.

Elizabeth Hazen brought her literary acumen and poetic intuition to careful editing of an early, imperfect draft. Her influence is felt throughout the published version.

Finally, at every stage of this book I have relied on the focused, constructive advice and unconditional support of my once and future coauthor, Margaret Hazen.

Notes

PROLOGUE
1. Sixteen of the presentations from the May 15–17, 2008, Deep Carbon Cycle Workshop can be viewed at Carnegie Institution for Science, "Sloan Deep Carbon Cycle Workshop—Sessions," accessed September 19, 2018, https://itunes.apple.com/us/podcast/sloan-deep-carbon-cycle-workshop -sessions/id438928309?mt=2.
2. Historical aspects of the Deep Carbon Observatory may be found on the organization's website, accessed October 12, 2018, http://deepcarbon.net.
3. The phone conversation with Jesse Ausubel took place in mid-January 2016, while I was on a writing retreat in Maui. The draft book manuscript was completed on Maui two years later.

MOVEMENT I—EARTH
1. For an account of Big Bang nucleosynthesis, see Carlos A. Bertulani, *Nuclei in the Cosmos* (Singapore: World Scientific, 2013).
2. See Fabio Iocco et al., "Primordial Nucleosynthesis: From Precision Cosmology to Fundamental Physics," *Physics Reports* 472 (2008): 1–76.
3. Carl Sagan, *Cosmos* (New York: Random House, 2002).
4. Quoted in Lindsay Smith, "Williamina Paton Fleming," *Project Continua* 1 (2015).
5. For an engaging history of the Harvard computers and astronomy of the late nineteenth and early twentieth centuries, see Dava Sobel, *The Glass Universe: How the Ladies of the Harvard Observatory Took the Measure of the Stars* (New York: Viking, 2016).

6. Quoted in Helen Fitzgerald, "Counted the Stars in the Heavens," *Brooklyn Daily Eagle*, September 18, 1927.

7. Quoted in J. Turner, "Cecilia Helena Payne-Gaposchkin," in *Contributions of 20th Century Women to Physics* (Los Angeles: UCLA Press, 2001).

8. Simon Mitton, *Fred Hoyle: A Life in Science* (New York: Cambridge University Press, 2011).

9. For a more detailed description, see D. A. Ostlie and B. W. Carroll, *An Introduction to Modern Stellar Astrophysics* (San Francisco: Addison-Wesley, 2007).

10. As quoted in Mitton, *Fred Hoyle*.

11. The timing of the first generation of stars is a matter of debate, but observations of distant galaxies, whose light began traveling less than a billion years after the Big Bang, indicates that large stars formed early in cosmic history. See D. P. Marrone et al., "Galaxy Growth in a Massive Halo in the First Billion Years of Cosmic History," *Nature* 553 (2018): 51–54.

12. The role of colliding neutron stars in the production of approximately half of the elements of the periodic table is described in D. Kasen et al., "Origin of the Heavy Elements in Binary Neutron-Star Mergers from a Gravitational-Wave Event," *Nature* 551 (2017): 80–84.

13. A useful overview of cosmochemistry is provided in Harry McSween and Gary Huss, *Cosmochemistry* (New York: Cambridge University Press, 2010).

14. The realization that diamond was likely the first mineral species is described in Robert M. Hazen et al., "Mineral Evolution," *American Mineralogist* 93 (2008): 1693–720.

15. The structures and properties of these crystalline forms of carbon are discussed in Robert M. Hazen, *The Diamond Makers* (New York: Cambridge University Press, 1999).

16. For a review of types and origins of meteorites, see James J. Papike, ed., *Planetary Materials* (Chantilly, VA: Mineralogical Society of America, 1998).

17. The known carbon minerals on Earth are surveyed in Robert M. Hazen et al., "The Mineralogy and Crystal Chemistry of Carbon," in *Carbon in Earth*, ed. Robert M. Hazen, Adrian P. Jones, and John Baross (Washington, DC: Mineralogical Society of America, 2013), 7–46. A complete up-to-date list of all carbon-bearing minerals is available at RRUFF Project, "IMA Database of Mineral Properties," accessed September 19, 2018, http://rruff.info/ima; while lists of their localities are at Mindat.org, accessed September 19, 2018, http://mindat.org.

18. Estimates of the total amount of crustal carbonate minerals come from Paul Falkowski et al., "The Global Carbon Cycle: A Test of Our Knowledge of Earth as a System," *Science* 290, no. 5490 (2000): 291–96; and Marc M. Hirschmann and Rajdeep Dasgupta, "The H/C Ratios of Earth's Near-

Surface and Deep Reservoirs, and Consequences for Deep Earth Volatile Cycles," *Chemical Geology* 262 (2009): 4–16.

19. For a review of early geological debates, see Martin J. S. Rudwick, *The Meaning of Fossils: Episodes in the History of Paleontology*, 2nd ed. (Chicago: University of Chicago Press, 1976).

20. For a biography of Hutton, see Jack Repcheck, *The Man Who Found Time: James Hutton and the Discovery of Earth's Antiquity* (New York: Perseus, 2003). Hutton's ideas gained traction following the publication of Charles Lyell, *Principles of Geology: Being an Attempt to Explain the Former Changes of the Earth's Surface, by Reference to Causes Now in Operation*, 3 vols. (London: Murray, 1830–33).

21. James Hutton, *Theory of the Earth, with Proofs and Illustrations, in Four Parts*, 2 vols. (Edinburgh: Creech, 1795).

22. Hall's research is described in Simon Mitton, *Carbon from Crust to Core: A Chronicle of Deep Carbon Science* (New York: Cambridge University Press, forthcoming).

23. James Hall, "Account of a Series of Experiments, Shewing the Effects of Compression in Modifying the Action of Heat," *Transactions of the Royal Society of Edinburgh* 6 (1812): 75.

24. Hall, "Account of a Series of Experiments," 81.

25. See Barcode of Life, accessed September 17, 2018, https://phe.rockefeller.edu/barcode.

26. Mark Y. Stoeckle et al., "Commercial Teas Highlight Plant DNA Barcode Identification Successes and Obstacles," *Scientific Reports* 1 (2011): art. 42.

27. John Schwartz, "Fish Tale Has DNA Hook: Students Find Bad Labels," *New York Times*, August 21, 2008, A1.

28. Jesse Ausubel, as quoted in Schwartz, "Fish Tale Has DNA Hook."

29. Our results are published in Robert M. Hazen and Jesse Ausubel, "On the Nature and Significance of Rarity in Mineralogy," *American Mineralogist* 101 (2016): 1245–51, https://doi.org/10.2138/am-2016-5601CCBY. We found that mineral rarity holds several parallels to biological rarity, as considered in Deborah Rabinowitz, "Seven Forms of Rarity," in *The Biological Aspects of Rare Plant Conservation*, ed. J. Synge (New York: Wiley, 1981), 205–17.

30. Sean Solomon, director of the Lamont-Doherty Earth Observatory of Columbia University, in remarks at Robert M. Hazen, "Mineralogical Co-evolution of the Geo- and Biospheres: Metallogenesis, the Supercontinent Cycle, and the Rise of the Terrestrial Biosphere" (Arthur D. Storke Lecture, Lamont-Doherty Earth Observatory, October 11, 2013).

31. The rare mineral fingerite was described in John M. Hughes and Chris G. Hadidiacos, "Fingerite, $Cu_{11}O_2(VO_4)_6$, a New Vanadium Sublimate from

Izalco Volcano, El Salvador: Descriptive Mineralogy," *American Mineralogist* 70 (1985): 193–96.

32. Information on our ongoing research in data-driven discovery can be found at Carnegie Science DTDI, accessed September 19, 2018, http://dtdi.carnegiescience.edu.

33. Biographical information on Robert Downs was obtained through interviews and emails with him in January and April 2017 and January 2018.

34. See Robert T. Downs, "The RRUFF Project: An Integrated Study of the Chemistry, Crystallography, Raman and Infrared Spectroscopy of Minerals," in *Program & Abstracts: 19th General Meeting of the International Mineralogical Association, Kobe, Japan, July 23–28, 2006* (Kobe: IMA, 2006), 3–13.

35. Biographical information on Jolyon Ralph was obtained through interviews and emails with him in August 2017 and January 2018.

36. In 2014, Jolyon Ralph donated the entire Mindat.org website and database to the not-for-profit Hudson Institute of Mineralogy so that it can be protected and continue to be freely available to all.

37. See Roberta L. Rudnick and S. Gao, "Composition of the Continental Crust," in *The Crust: Treatise on Geochemistry*, ed. Roberta L. Rudnick (New York: Elsevier, 2005), 1–64.

38. See B. J. McGill et al., "Species Abundance Distributions: Moving beyond Single Prediction Theories to Integration within an Ecological Framework," *Ecological Letters* 10 (2007): 995–1015. They state that the hollow, right-skewed curve for the frequency distribution of species seems to be a universal law in biology and ecology.

39. Mathematical approaches to lexical statistics are summarized in R. H. Baayen, *Word Frequency Distributions* (New York: Kluwer, 2001).

40. Biographical information on Grethe Hystad was obtained through interviews and emails with her in February 2017.

41. The original description of "mineral ecology" and application of LNRE formalisms to mineral distributions appears in Robert M. Hazen et al., "Mineral Ecology: Chance and Necessity in the Mineral Diversity of Terrestrial Planets," *Canadian Mineralogist* 53 (2015): 295–323.

42. Three papers published in 2015 are Grethe Hystad, Robert T. Downs, and Robert M. Hazen, "Mineral Frequency Distribution Data Conform to a LNRE Model: Prediction of Earth's 'Missing' Minerals," *Mathematical Geosciences* 47 (2015): 647–61; Robert M. Hazen et al., "Earth's 'Missing' Minerals," *American Mineralogist* 100 (2015): 2344–47; and Grethe Hystad et al., "Statistical Analysis of Mineral Diversity and Distribution: Earth's Mineralogy Is Unique," *Earth and Planetary Science Letters* 426 (2015): 154–57. See also the historical description of the development of mineral ecology at

Carnegie Science, "About Dr. Hazen," accessed September 19, 2018, http://hazen.carnegiescience.edu.

43. Subsequent mineral ecology papers considered, in turn, boron: Edward S. Grew et al., "How Many Boron Minerals Occur in Earth's Upper Crust?" *American Mineralogist* 102 (2017): 1573–87; chromium: Chao Liu et al., "Chromium Mineral Ecology," *American Mineralogist* 102 (2017): 612–19; and cobalt: Robert M. Hazen et al., "Cobalt Mineral Ecology," *American Mineralogist* 102 (2017): 108–16.

44. See Robert M. Hazen et al., "Carbon Mineral Ecology: Predicting the Undiscovered Minerals of Carbon," *American Mineralogist* 101 (2016): 889–906.

45. See Carbon Mineral Challenge, accessed September 19, 2018, http://mineralchallenge.net.

46. Biographical information on Dan Hummer was obtained through interviews and emails with him in July, August, and December 2017.

47. A mineral description is available in I. V. Pekov et al., "Tinnunculite, $C_5H_4N_4O_3 \cdot 2H_2O$: Finds at Kola Peninsula, Redefinition and Validation as a Mineral Species," *Zapiski Rossiiskogo Mineralogicheskogo Obshchetstva* 145, no. 4 (2016): 20–35.

48. High-pressure carbon mineralogy is reviewed in Artem Oganov et al., "Deep Carbon Mineralogy," in *Carbon in Earth*, ed. Robert M. Hazen, Adrian P. Jones, and John Baross (Washington, DC: Mineralogical Society of America, 2013), 44–77.

49. Merrill and Bassett's high-pressure calcite studies are found in Leo Merrill and William A. Bassett, "The Crystal Structure of $CaCO_3(II)$, a High-Pressure Metastable Phase of Calcium Carbonate," *Acta Crystallographica* B31 (1975): 343–49.

50. A history of the development and use of the diamond-anvil cell appears in Hazen, *Diamond Makers*.

51. Leo Merrill and William A. Bassett, "Miniature Diamond Anvil Pressure Cell for Single Crystal X-Ray Diffraction Studies," *Review of Scientific Instruments* 45 (1974): 290–94.

52. See Oganov, "Deep Carbon Mineralogy," for a review.

53. Biographical information on Marco Merlini was obtained through interviews and emails with him in August and September 2017, and a laboratory visit to Milan in May 2013.

54. Marco Merlini et al., "Structures of Dolomite at Ultrahigh Pressure and Their Influence on the Deep Carbon Cycle," *Proceedings of the National Academy of Sciences USA* 109 (2012): 13509–14.

55. Marco Merlini et al., "The Crystal Structures of $Mg_2Fe_2C_4O_{13}$, with Tetrahedrally Coordinated Carbon and $Fe_{13}O_{19}$, Synthesized at Deep Man-

tle Conditions," *American Mineralogist* 100 (2015): 2001–4. See also Marco Merlini et al., "Dolomite-IV: Candidate Structure for a Carbonate in the Earth's Lower Mantle," *American Mineralogist* 102 (2017): 1763–66.

56. A comprehensive overview of diamond research is found in Stephen B. Shirey et al., "Diamonds and the Geology of Mantle Carbon," *Reviews in Mineralogy and Geochemistry* 75 (2013): 355–421.

57. Evan M. Smith et al., "Large Gem Diamonds from Metallic Liquid in Earth's Deep Mantle," *Science* 354, no. 6318 (2016): 1403–5.

58. Overviews of diamond inclusion research include Steven B. Shirey and Stephen H. Richardson, "Start of the Wilson Cycle at 3 Ga Shown by Diamonds from Subcontinental Mantle," *Science* 333 (2011): 434–36. See also Shirey et al., "Diamonds and the Geology of Mantle Carbon."

59. See Gemological Institute of America, accessed September 19, 2018, https://www.gia.edu.

60. The finding that diamond inclusions are different in gems older than 3 billion years was reported by Shirey and Richardson, "Start of the Wilson Cycle at 3 Ga."

61. Biographical information on Francis Birch is provided in Thomas J. Ahrens, *Albert Francis Birch, 1903–1992* (Washington, DC: National Academy of Sciences, 1998).

62. Francis Birch, "Elasticity and Constitution of the Earth's Interior," *Journal of Geophysical Research* 57 (1952): 227–86.

63. From Birch, "Elasticity and Constitution," 234.

64. Isotopic evidence for bulk Earth fractionation and carbon in the core is reviewed in Bernard J. Wood, Jei Li, and Anat Shahar, "Carbon in the Core: Its Influence on the Properties of Core and Mantle," *Reviews in Mineralogy and Geochemistry* 75 (2013): 231–50. See also Anat Shahar et al., "High-Temperature Si Isotope Fractionation between Iron Metal and Silicate," *Geochimica et Cosmochimica Acta* 75 (2011): 7688–97.

65. Biographical information on Jie Li was obtained through interviews and emails with her in June 2017.

66. Bin Chen et al., "Hidden Carbon in Earth's Inner Core Revealed by Shear Softening in Dense Fe_7C_3," *Proceedings of the National Academy of Sciences USA* 111 (2014): 17755–58.

67. Clemens Prescher et al., "High Poisson's Ratio of Earth's Inner Core Explained by Carbon Alloying," *Nature Geoscience* 8 (2015): 220–23.

68. Prescher et al., "High Poisson's Ratio."

69. Hystad, "Statistical Analysis of Mineral Diversity."

70. This section is adapted from Robert M. Hazen, "Mineral Fodder," *Aeon*, June 24, 2014, https://aeon.co/essays/how-life-made-the-earth-into-a-cosmic-marvel.

MOVEMENT II—AIR

1. A portrait of Earth's formation from the solar nebula appears in Robert M. Hazen, *The Story of Earth: The First 4.5 Billion Years, from Stardust to Living Planet* (New York: Viking, 2012).

2. The average compositions of the chondrite meteorites that formed Earth are summarized in H. Palme, K. Lodders, and A. Jones, "Solar System Abundances of the Elements," *Treatise on Geochemistry* 2 (2014): 15–35.

3. For a detailed description of organic molecules found in meteorites, see Mark A. Sephton, "Organic Compounds in Carbonaceous Meteorites," *Natural Products Report* 19 (2002): 292–311. For a useful review, see Puna Dalai, Hussein Kaddour, and Nita Sahai, "Incubating Life: Prebiotic Sources of Organics for the Origin of Life," *Elements* 12 (2016): 401–6.

4. For a summary of speculations regarding the earliest atmosphere, see Kevin Zahnle, "Earth's Earliest Atmosphere," *Elements* 2 (2006): 217–22.

5. For a recent theory of the Moon's formation with references to earlier ideas, see Matija Ćuk et al., "Tidal Evolution of the Moon from a High-Obliquity, High-Angular-Momentum Earth," *Nature* 539 (2016): 402–6. See also Hazen, *Story of Earth*.

6. The faint young Sun idea was introduced in Carl Sagan and George Mullen, "Earth and Mars: Evolution of Atmospheres and Surface Temperatures," *Science* 177 (1972): 52–56.

7. See I. Rasool and C. De Bergh, "The Runaway Greenhouse and the Accumulation of CO_2 in the Venus Atmosphere," *Nature* 226, no. 5250 (1970): 1037–39.

8. See, for example, John C. Armstrong, L. E. Wells, and G. Gonzales, "Rummaging through Earth's Attic for Remains of Ancient Life," *Icarus* 160 (2002): 183–96.

9. This DCO field trip, led by Italian geologists Carlo Cardellini, Giovanni Chiodini, Matteo Lelli, and Stefano Caliro, occurred on Tuesday, October 6, 2015.

10. A typical example is found in James S. Trefil and Robert M. Hazen, *The Sciences: An Integrated Approach*, 8th ed. (Hoboken, NJ: Wiley, 2015), 431.

11. The deep carbon cycle has been the central focus of the "Reservoirs and Fluxes" community of the DCO. See Deep Carbon Observatory, "Reservoirs and Fluxes," accessed September 21, 2018, https://deepcarbon.net/content/reservoirs-and-fluxes.

12. See, for example, Peter B. Kelemen and Craig E. Manning, "Reevaluating Carbon Fluxes: What Goes Down, Mostly Comes Up," *Proceedings of the National Academy of Sciences USA* 112 (2015): E3997-4006.

13. The rise of planktonic calcifiers and the implications of a mid-Mesozoic

revolution are described in Andy Ridgwell, "A Mid-Mesozoic Revolution in the Regulation of Ocean Chemistry," *Marine Geology* 217 (2005): 339–57. See also Andy Ridgwell and Richard E. Zeebe, "The Role of the Global Carbonate Cycle in the Regulation and Evolution of the Earth System," *Earth and Planetary Science Letters* 234 (2005): 299–315.

14. Terry Plank, interview by email, January 10, 2018.

15. Sverjensky gave his short talk during the Sloan Deep Carbon Cycle Workshop in Washington, DC, on May 15 and 16, 2008. Many talks from the workshop appear at Carnegie Institution for Science, "Sloan Deep Carbon Cycle Workshop—Sessions," accessed September 19, 2018, https://itunes.apple.com/us/podcast/sloan-deep-carbon-cycle-workshop-sessions/id438928309?mt=2.

16. Ding Pan et al., "Dielectric Properties of Water under Extreme Conditions and Transport of Carbon in the Deep Earth," *Proceedings of the National Academy of Sciences USA* 110 (2013): 6646–50.

17. Studies of carbonate solubilities appear in S. Facq et al., "*In situ* Raman Study and Thermodynamic Model of Aqueous Carbonate Speciation in Equilibrium with Aragonite under Subduction Zone Conditions," *Geochimica et Cosmochimica Acta* 132 (2014): 375–90. Additional information was provided by email interview with Isabelle Daniel on January 12, 2018.

18. Biographical information on Dimitri Sverjensky was obtained through email interviews with him on January 11, 12, and 14, 2018.

19. A few of our publications on mineral surfaces include Christine M. Jonsson et al., "Attachment of L-Glutamate to Rutile (α-TiO_2): A Potentiometric, Adsorption and Surface Complexation Study," *Langmuir* 25 (2009): 12127–35; Namhey Lee et al., "Speciation of L-DOPA on Nanorutile as a Function of pH and Surface Coverage Using Surface-Enhanced Raman Spectroscopy (SERS)," *Langmuir* 28 (2012): 17322–30; Charlene Estrada et al., "Interaction between L-Aspartate and the Brucite [$Mg(OH)_2$]–Water Interface," *Geochimica et Cosmochimica Acta* 155 (2015): 172–86; and Teresa Fornaro et al., "Binding of Nucleic Acid Components to the Serpentinite-Hosted Hydrothermal Mineral Brucite," *Astrobiology* 18, no. 8 (August 2018): 989–1007, https://doi.org/10.1089/ast.2017.1784.

20. Dimitri A. Sverjensky, Brandon Harrison, and David Azzolini, "Water in the Deep Earth: The Dielectric Constant and the Solubilities of Quartz and Corundum to 60 kb and 1,200 °C," *Geochimica et Cosmochimica Acta* 129 (2014): 125–45.

21. See Fang Huang et al., "Immiscible Hydrocarbon Fluids in the Deep Carbon Cycle," *Nature Communications* 8 (2017): art. 15798.

22. Dimitri A. Sverjensky and Fang Huang, "Diamond Formation Due to a

pH Drop during Fluid-Rock Interactions," *Nature Communications* 6 (2015): art. 8702.

23. Background information on deep methane and methane isotopologues is found in Mark A. Sephton and Robert M. Hazen, "On the Origins of Deep Hydrocarbons," *Reviews in Mineralogy and Geochemistry* 75 (2013): 449–65.

24. John M. Eiler and Edwin Schauble, "$^{18}O^{13}C^{16}O$ in Earth's Atmosphere," *Geochimica et Cosmochimica Acta* 68 (2004): 4767–77.

25. Biographical information on Edward Young was obtained through emails with him on January 10, 2018.

26. The development of Panorama is described in Edward D. Young et al., "A Large-Radius High-Mass-Resolution Multiple-Collector Isotope Ratio Mass Spectrometer for Analysis of Rare Isotopologues of O_2, N_2, CH_4 and Other Gases," *International Journal of Mass Spectrometry* 401 (2016): 1–10. Early applications were published in Edward D. Young et al., "The Relative Abundances of Resolved $^{12}CH_2D_2$ and $^{13}CH_3D$ and Mechanisms Controlling Isotopic Bond Ordering in Abiotic and Biotic Methane Gases," *Geochimica et Cosmochimica Acta* 203 (2017): 235–64.

27. Biographical information on Shuhei Ono was obtained through emails with him on January 10, 2018. Ono's developments are described in Shuhei Ono et al., "Measurement of a Doubly-Substituted Methane Isotopologue, $^{13}CH_3D$, by Tunable Infrared Laser Direct Absorption Spectroscopy," *Analytical Chemistry* 86 (2014): 6487–94.

28. Subsequent papers include A. R. Whitehill et al., "Clumped Isotope Effects during OH and Cl Oxidation of Methane," *Geochimica et Cosmochimica Acta* 196 (2017): 307–25; and D. T. Wang et al., "Clumped Isotopologue Constraints on the Origin of Methane at Seafloor Hot Springs," *Geochimica et Cosmochimica Acta* 223 (2018): 141–58.

29. Estimates of carbon fluxes from major volcanoes are found in Michel R. Burton, Georgina M. Sawyer, and Dominico Granieri, "Deep Carbon Emissions from Volcanoes," *Reviews in Mineralogy and Geochemistry* 75 (2013): 323–54.

30. The activities and numerous publications of DECADE scientists are found at Deep Carbon Observatory, "DECADE," accessed September 22, 2018, https://deepcarboncycle.org/home-decade.

31. See NOVAC, "Project Partners," accessed June 27, 2018, http://www.novac-project.eu/partners.htm.

32. The use of sulfur dioxide and carbon dioxide in tandem to monitor volcanoes is described in A. Aiuppa et al., "Forecasting Etna Eruptions by Real-Time Observation of Volcanic Gas Composition," *Geology* 35 (2007): 1115–18. For an application of the method, see J. M. de Moor et al., "Turmoil at Turrialba Volcano (Costa Rica): Degassing and Eruptive Processes

Inferred from High-Frequency Gas Monitoring," *Journal of Geophysical Research—Solid Earth* 121, no. 8 (2016): 5761–75.

33. See Peter W. Lipman and Donal R. Mullineaux, eds., dedication in *The 1980 Eruptions of Mount Saint Helens, Washington*, Geological Survey Professional Paper 1250 (Washington, DC: US Government Printing Office, 1981), vii.

34. See R. V. Fisher, "Obituary Harry Glicken (1958–1991)," *Bulletin of Volcanology* 53, no. 6 (1991): 514–16.

35. For differing descriptions of the events leading to multiple deaths on January 14, 1993, see Stanley Williams and Fen Montaigne, *Surviving Galeras* (New York: Houghton Mifflin, 2001); and Victoria Bruce, *No Apparent Danger: The True Story of Volcanic Disaster at Galeras and Nevada del Ruiz* (New York: HarperCollins, 2002).

36. Biographical information on Marie Edmonds was obtained through emails with her on January 12 and 16, 2018.

37. Emily Mason, Marie Edmonds, and Alexandra V. Turchyn, "Remobilization of Crustal Carbon May Dominate Volcanic Arc Emissions," *Science* 357, no. 6346 (2017): 290–94.

38. For a list of Earth's volcanoes, see Smithsonian Institution, Global Volcanism Program, accessed June 27, 2018, https://volcano.si.edu.

39. Fluid inclusions in diamonds are discussed in Steven B. Shirey et al., "Diamonds and the Geology of Mantle Carbon," *Reviews in Mineralogy and Geochemistry* 75 (2013): 355–421.

40. See Shirey et al., "Diamonds and the Geology of Mantle Carbon." Additional background on DCO diamond research was provided by Steven Shirey in emails on January 12, 2018.

41. Kelemen and Manning, "Reevaluating Carbon Fluxes." Alternative values of deep carbon reservoirs and fluxes were provided in Rajdeep Dasgupta and Marc M. Hirschmann, "The Deep Carbon Cycle and Melting in Earth's Interior," *Earth and Planetary Science Letters (Frontiers)* 298 (2010): 1–13.

42. Estimates of historic fossil fuel consumption are found in documents of the US Energy Information Administration, "History of Energy Consumption in the United States, 1775–2009," February 9, 2011, https://www.eia.gov/todayinenergy/detail.php?id=10.

43. The literature on greenhouse gases and global warming is vast and overwhelming. Key reports include Gabriele C. Hegerl et al., "Understanding and Attributing Climate Change," in *Contribution of Working Group I to the Fourth Assessment Report of the Intergovernmental Panel on Climate Change, 2007*, ed. S. Solomon et al. (Cambridge: Cambridge University Press, 2007), chap. 9; National Research Council, *Advancing the Science of Climate Change* (Washington, DC: National Academies Press, 2010); and Intergov-

ernmental Panel on Climate Change, *Fifth Assessment Report*, 4 vols. (New York: Cambridge University Press, 2013).

44. The story was reported in Kendra Pierre-Louis and Nadja Popovich, "Of 21 Winter Olympic Cities, Many May Soon Be Too Warm to Host the Games," *New York Times*, January 11, 2018.

45. For positive feedback in the context of methane release, see D. M. Lawrence and A. Slater, "A Projection of Severe Near-Surface Permafrost Degradation during the 21st Century," *Geophysical Research Letters* 32, no. 24 (2005): L24401; David Archer, "Methane Hydrate Stability and Anthropogenic Climate Change," *Biogeosciences* 4, no. 4 (2007): 521–44.

46. For more information, see Oman Drilling Project, accessed June 27, 2018, http://www.omandrilling.ac.uk.

47. This section draws on Jesse H. Ausubel, "A Census of Ocean Life: On the Difficulty and Joy of Seeing What Is Near and Far," *SGI Quarterly* 60 (April 2010): 6–8.

MOVEMENT III—FIRE

1. The history of scoring in American-style football is reviewed in David M. Nelson, *The Anatomy of a Game* (Newark: University of Delaware Press, 1994).

2. For an overview of organic chemistry, see Marye Anne Fox and James K. Whitesell, *Organic Chemistry*, 3rd ed. (Sudbury, MA: Jones and Bartlett, 2004).

3. Alasdair H. Neilson, ed., *PAHs and Related Compounds: Chemistry* (Berlin: Springer, 1998). See also Chunshan Song, *Chemistry of Diesel Fuels* (Boca Raton, FL: CRC Press, 2015).

4. Titan's rivers and lakes are described in E. R. Stofan et al., "The Lakes of Titan," *Nature* 445 (2007): 61–64. See also A. Coustenis and F. W. Taylor, *Titan: Exploring an Earthlike World* (Singapore: World Scientific, 2008).

5. For a review of petroleum refining, see James G. Speight, *The Chemistry and Technology of Petroleum*, 4th ed. (New York: Marcel Dekker, 2006).

6. See, for example, A. I. Railkin, *Marine Biofouling: Colonization Processes and Defenses* (Boca Raton, FL: CRC Press, 2004); and Laurel Hamers, "Designing a Better Glue from Slug Goo," *Science News*, September 30, 2017, 14–15. For an interesting approach to antifouling, see Shahrouz Amini et al., "Preventing Mussel Adhesion Using Lubricant-Infused Materials," *Science* 357, no. 6352 (2017): 668–72.

7. See Edward M. Petrie, *Handbook of Adhesives and Sealants* (New York: McGraw-Hill, 2000).

8. The world premiere took place on February 10, 1975, in Harvard's historic Sanders Theatre. We made a recording at a studio the next day at Interme-

dia Sound Studios, 331 Newbury Street, Boston. However, nearby construction created intermittent background noises that made the recording unsuitable for release. Additional details came from email interviews with Martha Kim on September 27 and October 1, 2017.

9.   Geim and Novoselov's Nobel Prize–winning research was published in K. S. Novoselov et al., "Electric Field Effect in Atomically Thin Carbon Films," *Science* 306 (2004): 666–69. See also Andre K. Geim and Konstantin S. Novoselov, "The Rise of Graphene," *Nature Materials* 6 (2007): 183–91.

10. Andre K. Geim and Phillip Kim, "Carbon Wonderland," *Scientific American* 298 (April 2008): 90–97; Edward L. Wolf, *Applications of Graphene: An Overview* (Berlin: Springer, 2014).

11. Mitch Jacoby, "Graphene Finds New Use as Hair Dye," *Chemical and Engineering News*, March 10, 2018, 4.

12. Sumio Iijima, "Helical Microtubules of Graphitic Carbon," *Nature* 354 (1991): 56–58.

13. Varieties of carbon nanoarchitecture are reviewed in Peter J. F. Harris, *Carbon Nanotube Science: Synthesis, Properties and Applications* (New York: Cambridge University Press, 2009).

14. H. W. Kroto et al., "C_{60}: Buckminsterfullerene," *Nature* 318, no. 6042 (1985): 162–63. See also Richard E. Smalley, "Discovering the Fullerenes," *Reviews of Modern Physics* 69 (1997): 723–30.

15. See, for example, Guillaume Povie et al., "Synthesis of a Carbon Nanobelt," *Science* 356, no. 6334 (2017): 172–73.

16. As quoted at Wikiquote, s.v. *"The Graduate,"* accessed January 21, 2018, https://en.wikiquote.org/wiki/The_Graduate.

17. The history of rubber manufacturing is presented in Howard Wolf and Ralph Wolf, *Rubber: A Story of Glory and Greed* (Akron, OH: Smithers Rapra, 2009).

18. Hermann Staudinger, "Über Polymerisation," *Berichte der Deutschen Chemischen Gesellschaft* 53, no. 6 (1920): 1073–85.

19. See Mary Ellen Bowden, *Chemical Achievers: The Human Face of the Chemical Sciences* (Philadelphia: Chemical Heritage Foundation, 1997). See also Jeffrey L. Meikle, *American Plastics: A Cultural History* (New Brunswick, NJ: Rutgers University Press, 1997).

20. The life and work of Wallace Carothers are reviewed in Meikle, *American Plastics*.

21. The polyurethane demonstration kit is sold by Flinn Scientific (Batavia, Illinois) as the "Polyurethane Foam System—Chemical Demonstration Kit," item#C0335, accessed April 22, 2018, https://www.flinnsci.com/polyurethane-foam-system---chemical-demonstration-kit/c0335.

22. A useful review of the genetic mutation that causes sickle cell anemia is

provided in D. C. Rees, T. N. Williams, and M. T. Gladwin, "Sickle Cell Disease," *Lancet* 376, no. 9757 (2010): 2018–31.

23. See Aamer Ali Shah et al., "Biological Degradation of Plastics: A Comprehensive Review," *Biotechnology Advances* 26 (2008): 246–65.

24. Nylon depolymerization is discussed in M. Moezzi and M. Ghane, "The Effect of UV Degradation on Toughness of Nylon 66/Polyester Woven Fabrics," *Journal of the Textile Institute* 104, no. 12 (2013): 1277–83.

25. See Marcella Hazan, *Essentials of Classic Italian Cooking* (New York: Knopf, 1992). Information on the preparation of artisanal Italian pasta came in part from in-person interviews with Teresa Fornaro in May of 2018.

26. See Library of Congress, "The Deterioration and Preservation of Paper: Some Essential Facts," accessed March 15, 2018, https://www.loc.gov/preservation /care/deterioratebrochure.html.

MOVEMENT IV—WATER

1. Among the many books on the philosophy of the quest for life's origins, see Iris Fry, *The Emergence of Life on Earth: A Historical and Scientific Overview* (New Brunswick, NJ: Rutgers University Press, 2000); and Constance M. Bertka, ed., *Exploring the Origin, Extent, and Future of Life: Philosophical, Ethical and Theological Perspectives* (Washington, DC: American Association for the Advancement of Science, 2009).

2. The timing of the Moon's formation is reviewed in Tais W. Dahl and David J. Stevenson, "Turbulent Mixing of Metal and Silicate during Planet Accretion—an Interpretation of the Hf-W Chronometer," *Earth and Planetary Science Letters* 295, no. 1–2 (2010): 177–86. The age of Earth's oldest fossils is a matter of debate, but most paleontologists accept 3.5-billion-year-old stromatolites from Western Australia: Abigail Allwood et al., "Stromatolite Reef from the Early Archean of Australia," *Nature* 441, no. 7094 (2006): 714–18. Plausible 3.7-billion-year-old fossils have been described from Greenland: T. Hassenkam et al., "Elements of Eoarchean Life Trapped in Mineral Inclusions," *Nature* 548 (2017): 78–81. Even older signs of life have been proposed in Matthew S. Dodd et al., "Evidence for Early Life in Earth's Oldest Hydrothermal Vent Precipitates," *Nature* 543 (2017): 60–64; Takayuki Tashiro et al., "Early Trace of Life from 3.95 Ga Sedimentary Rocks in Labrador, Canada," *Nature* 549 (2017): 516–18.

3. The possibility that Earth's biosphere was seeded by Martian meteorites is examined in C. Mileikowsky et al., "Natural Transfer of Microbes in Space, Part I: From Mars to Earth and Earth to Mars," *Icarus* 145, no. 2 (2000): 391–427.

4. See Simon Mitton, *Fred Hoyle: A Life in Science* (New York: Cambridge University Press, 2011).

5. For more complete discussions of the definition of life and theories of life's origins, see Noam Lahav, *Biogenesis: Theories of Life's Origins* (New York: Oxford University Press, 1999); Fry, *Emergence of Life on Earth*; Robert M. Hazen, *Genesis: The Scientific Quest for Life's Origin* (Washington, DC: Joseph Henry Press, 2005); David Deamer and Jack W. Szostak, eds., *The Origins of Life* (Cold Spring Harbor, NY: Cold Spring Harbor Laboratory Press, 2010); Eric Smith and Harold J. Morowitz, *The Origin and Nature of Life on Earth: The Emergence of the Fourth Geosphere* (New York: Cambridge University Press, 2016).

6. The clay world hypothesis is presented in A. Graham Cairns-Smith, *Seven Clues to the Origin of Life* (Cambridge: Cambridge University Press, 1985); and A. Graham Cairns-Smith and Hyman Hartman, *Clay Minerals and the Origin of Life* (Cambridge: Cambridge University Press, 1986).

7. The original papers describing prebiotic organic synthesis are Stanley L. Miller, "A Production of Amino Acids under Possible Primitive Earth Conditions," *Science* 117 (1953): 528–29; and Stanley L. Miller, "Production of Some Organic Compounds under Possible Primitive Earth Conditions," *Journal of the American Chemical Society* 77 (1955): 2351–61. See also Christopher Wills and Jeffrey Bada, *The Spark of Life: Darwin and the Primeval Soup* (Cambridge, MA: Perseus, 2000).

8. For a short biography, see S. L. Miller and J. Oró, "Harold C. Urey 1893–1981," *Journal of Molecular Evolution* 17 (1981): 263–64.

9. Miller, "Production of Amino Acids."

10. Claude Lévi-Strauss, *La pensée sauvage* (Paris: Libraire Plon, 1962).

11. See Robert M. Hazen, "Deep Carbon and False Dichotomies," *Elements* 10 (2010): 407–9.

12. As quoted in Wills and Bada, *Spark of Life*, 41.

13. Miller's critique of the hydrothermal hypothesis is quoted in P. Radetsky, "How Did Life Start?" *Discover*, November 1992, 74–82. See also Hazen, *Genesis*, 109–110, 266 for additional context.

14. Paul M. Schenk et al., *Enceladus and the Icy Moons of Saturn* (Tucson: University of Arizona Press, 2018).

15. See David W. Deamer and R. M. Pashley, "Amphiphilic Components of the Murchison Carbonaceous Chondrite: Surface Properties and Membrane Formation," *Origins of Life and Evolution of the Biosphere* 19 (1989): 21–38. For a more general overview, see David W. Deamer, "Self-Assembly of Organic Molecules and the Origin of Cellular Life," *Reports of the National Center for Science Education* 23 (May–August 2003): 20–33.

16. Darwin's letter to Joseph Hooker is Item 7471 in the Darwin online database, accessed October 6, 2018, https://www.darwinproject.ac.uk/letter/DCP-LETT-7471.xml.

17. Biographical information on Charlene Estrada was obtained through interviews and emails with her in January 2017.

18. For Estrada's study on amino acid adsorption, see Charlene Estrada et al., "Interaction between L-Aspartate and the Brucite [Mg(OH)$_2$]–Water Interface," *Geochimica et Cosmochimica Acta* 155 (2015): 172–86.

19. Biographical information on Teresa Fornaro was obtained through interviews and emails with her in January 2017.

20. For Fornaro's study of nucleotides on brucite, see Teresa Fornaro et al., "Binding of Nucleic Acid Components to the Serpentinite-Hosted Hydrothermal Mineral Brucite," *Astrobiology* 18, no. 8 (August 2018): 989–1007, https://doi.org/10.1089/ast.2017.1784.

21. See Smith and Morowitz, *Origin and Nature of Life on Earth*, 186, 201, and following.

22. This section is adapted from Robert M. Hazen, "Chance, Necessity, and the Origins of Life," *Philosophical Transactions of the Royal Society. Series A* 375 (2016): 20160353.

23. Jacques Monod, *Chance and Necessity: An Essay on the Natural Philosophy of Modern Biology* (New York: Vintage Books, 1972). Monod's famous quote appears in the book's concluding paragraph.

24. Ernest Schoffeniels, *Anti-chance: A Reply to Monod's Chance and Necessity*, trans. B. L. Reid (Oxford: Pergamon, 1976), 18.

25. These ideas were first presented in a lecture at the Carnegie Institution: Robert Hazen, "Chance, Necessity, and the Origins of Life" (Carnegie Public Lectures, Carnegie Institution for Science, November 12, 2015), accessed October 12, 2018, https://carnegiescience.edu/events/lectures/special-event-robert-hazen-chance-necessity-and-origins-life.

26. Charles Darwin, *The Origin of Species* (London: John Murray, 1859).

27. See Paul G. Falkowski, *Life's Engines: How Microbes Made Earth Habitable* (Princeton, NJ: Princeton University Press, 2015). See also J. D. Kim et al., "Discovering the Electronic Circuit Diagram of Life: Structural Relationships among Transition Metal Binding Sites in Oxidoreductases," *Philosophical Transactions of the Royal Society. Series B* 368 (2013): 20120257.

28. Biographical information on Paul Falkowski comes from Falkowski, *Life's Engines*, 1–7. Additional information was obtained through email interviews with him on June 9, 2017, and December 10, 2017, and from an unpublished biosketch dated June 11, 2017.

29. Benjamin I. Jelen, Donato Giovannelli, and Paul G. Falkowski, "The Role of Microbial Electron Transfer in the Coevolution of the Geosphere and Biosphere," *Annual Review of Microbiology* 70 (2016): 45–62.

30. The study of deep microbial life is the focus of the DCO's Deep Life Com-

munity: Deep Carbon Observatory, "Deep Life," accessed October 5, 2018, https://deepcarbon.net/content/deep-life.

31. Census of Deep Life, accessed October 6, 2018, https://deepcarbon.net/tag/census-deep-life.

32. Biographical information on Steven D'Hondt was obtained through email interviews with him on December 20, 2017.

33. See Anurag Sharma et al., "Microbial Activity at Gigapascal Pressures," *Science* 295 (2002): 1514–16.

34. For a discussion of photosynthesis, see Falkowski, *Life's Engines*, 96 and following.

35. A recent study described organisms with modified chlorophyll that can harvest near-infrared radiation: Dennis J. Nürnberg et al., "Photochemistry beyond the Red Limit in Chlorophyll f–Containing Photosystems," *Science* 360 (2018): 1210–13.

36. Robert M. Hazen et al., "Mineral Evolution," *American Mineralogist* 93 (2008): 1693–1720.

37. The origins of multicellularity are described in Andrew H. Knoll, *Life on a Young Planet: The First Three Billion Years of Evolution on Earth* (Princeton, NJ: Princeton University Press, 2003), 161–78.

38. Knoll, *Life on a Young Planet*, 122–60.

39. Biographical information on Lynn Margulis was obtained through interviews with her in Amherst, Massachusetts, on November 3 and 4, 2011, just three weeks before her death.

40. Lynn Sagan, "On the Origin of Mitosing Cells," *Journal of Theoretical Biology* 14 (1967): 225–74. See also Lynn Margulis, *Origin of Eukaryotic Cells* (New Haven, CT: Yale University Press, 1970).

41. As quoted in Charles Mann, "Lynn Margulis: Science's Unruly Earth Mother," *Science* 252 (April 19, 1991): 379–81.

42. Ediacaran fossils are reviewed in Knoll, *Life on a Young Planet*, 164–78. Additional information was provided by Drew Muscente in an email dated June 10, 2017.

43. Biographical information on Michael Meyer was obtained through conversations with him in the spring of 2017 and an email interview on June 7, 2017.

44. Paleobiology Database, accessed September 29, 2018, https://paleobiodb.org.

45. The Ediacaran study, "Deep-Time Data-Driven Discovery and the Co-evolution of the Geosphere and Biosphere," was presented as part of a lecture by Robert Hazen et al. at the National Science Foundation, Arlington, Virginia, on May 4, 2017.

46. The use of network analysis to identify mass extinctions in the fossil record appears in A. Drew Muscente et al., "Quantifying Ecological Impacts of

Mass Extinctions with Network Analysis of Fossil Communities," *Proceedings of the National Academy of Sciences USA* 115 (2018): 5217–22.

47. Key references on biomineralization are cited in Patricia Dove, "The Rise of Skeletal Biominerals," *Elements* 6, no. 1 (2010): 37–42. Insights on coral formation are in Stanislas Von Euw et al., "Biological Control of Aragonite Formation in Stony Corals," *Science* 356, no. 6341 (2017): 933–38.

48. Biographical information on Patricia Dove was obtained through email interviews with her on June 23, 2017, and January 11, 2018.

49. See S. Weiner et al., "Biologically Formed Amorphous Calcium Carbonate," *Connective Tissue Research* 44 (2003): 214–18. See also D. Wang et al., "Carboxylated Molecules Regulate Magnesium Content of Amorphous Calcium Carbonates during Calcification," *Proceedings of the National Academy of Sciences USA* 106 (2009): 21511–16.

50. See Hans R. Thierstein and Jeremy R. Young, *Coccolithophores: From Molecular Processes to Global Impact* (Berlin: Springer, 2004).

51. The rise of the terrestrial biosphere is recounted in David Beerling, *The Emerald Planet: How Plants Changed Earth's History* (New York: Oxford University Press, 2007).

52. Heather M. Wilson and Lyall I. Anderson, "Morphology and Taxonomy of Paleozoic Millipedes (Diplopoda: Chilignatha: Archipolypoda) from Scotland," *Journal of Paleontology* 78 (2004): 169–84.

53. Biographical information on Neil Shubin was obtained through interviews and emails with him in January 2018.

54. The discovery is described in Edward B. Daeschler, Neil H. Shubin, and Farish A. Jenkins Jr., "A Devonian Tetrapod-like Fish and the Evolution of the Tetrapod Body Plan," *Nature* 440, no. 7085 (2006): 757–63. A popular account appears in Neil H. Shubin, *Your Inner Fish: A Journey into the 3.5-Billion-Year History of the Human Body* (New York: Vintage Books, 2008).

55. For a classic overview, see Dirk Willem van Krevelen, *Coal: Typology, Chemistry, Physics and Constitution*, 3rd ed. (New York: Elsevier Science, 1993).

56. The role of clay minerals in sequestering organic carbon is reviewed in Martin J. Kennedy and Thomas Wagner, "Clay Mineral Continental Amplifier for Marine Carbon Sequestration in a Greenhouse Ocean," *Proceedings of the National Academy of Sciences USA* 108 (2011): 9776–81.

57. For an account of radiocarbon dating and its applications, see R. E. Taylor, *Radiocarbon Dating: An Archeological Perspective* (Orlando, FL: Academic Press, 1987).

58. Taylor, *Radiocarbon Dating*, chap. 6.

59. Michael R. Waters et al., "Late Pleistocene Horse and Camel Hunting at the Southern Margin of the Ice-Free Corridor: Reassessing the Age of Wally's

Beach, Canada," *Proceedings of the National Academy of Sciences USA* 112, no. 14 (2015): 4263–67.

60. Quan Hua, Mike Barbetti, and Andrzej Z. Rakowski, "Atmospheric Radiocarbon for the Period 1950–2010," *Radiocarbon* 55 (2013): 2059–72.

61. For an entertaining alternate calculation of molecular cycling, see Sam Kean, *Caesar's Last Breath: Decoding the Secrets of the Air around Us* (New York: Little, Brown, 2017).

Index